Management and Industrial Engineering

Series Editor

J. Paulo Davim, Department of Mechanical Engineering, University of Aveiro, Aveiro, Portugal

This series fosters information exchange and discussion on management and industrial engineering and related aspects, namely global management, organizational development and change, strategic management, lean production, performance management, production management, quality engineering, maintenance management, productivity improvement, materials management, human resource management, workforce behavior, innovation and change, technological and organizational flexibility, self-directed work teams, knowledge management, organizational learning, learning organizations, entrepreneurship, sustainable management, etc. The series provides discussion and the exchange of information on principles, strategies, models, techniques, methodologies and applications of management and industrial engineering in the field of the different types of organizational activities. It aims to communicate the latest developments and thinking in what concerns the latest research activity relating to new organizational challenges and changes world-wide. Contributions to this book series are welcome on all subjects related with management and industrial engineering. To submit a proposal or request further information, please contact Professor J. Paulo Davim, Book Series Editor, pdavim@ua.pt

More information about this series at http://www.springer.com/series/11690

Waqas Nawaz · Muammer Koç

Industry, University and Government Partnerships for the Sustainable Development of Knowledge-Based Society

Drivers, Models and Examples in US, Norway, Singapore and Qatar

 Springer

Waqas Nawaz
Sustainable Development Division
Hamad Bin Khalifa University
Doha, Qatar

Muammer Koç
Sustainable Development Division
Hamad Bin Khalifa University
Doha, Qatar

ISSN 2365-0532   ISSN 2365-0540   (electronic)
Management and Industrial Engineering
ISBN 978-3-030-26801-5   ISBN 978-3-030-26799-5   (eBook)
https://doi.org/10.1007/978-3-030-26799-5

This Springer imprint is published by the registered company Springer Nature Switzerland AG
The registered company address is: Gewerbestrasse 11, 6330 Cham, Switzerland

Preface

Transformation to sustainable development can only be achieved through a truly diversified and innovation-driven knowledge economy. Although the concept of knowledge ecosystem is not new, the roles of industries, universities and government in innovation-driven knowledge-based economy have evolved over time. The growing interaction between the educational system; research, development and innovation capacity; and the public and private sector has led to a higher number and quality of economic opportunities, which were restricted to a fewer economic sectors and fields in the past. The importance and urgency of such transformation and critical role of developing functional partnerships between academia, industry and government have been even more strongly pronounced for countries relying on single or few resources such as oil, gas and minerals. Oil and gas rich countries have come to realize that the growth emanated from hydrocarbon economy is unsustainable and indeed insecure in the long run. Therefore, many oil and gas rich countries have been looking forward to the means of diversifying their economy through development of strategies, policies and implementation roadmaps for effective industry-university-government partnerships (IUGP), along with promoting entrepreneurship, private enterprises, public-private partnerships and relevant innovation capacity.

This book addresses the rapidly growing interest in economic diversification through partnerships between industry, university and government, with a focus on the economic diversification of the State of Qatar as an example for a country with rich natural gas resources and in transition to a knowledge economy in preparation for an envisioned sustainable development. The book provides a comparative account on the knowledge ecosystems in United States, Norway, Singapore, and Qatar. It brings forward an evolutionary national economic-transformational perspective around legislation, institutional and cultural settings, intermediary structures, and support programs.

The book adopts the modern Triple Helix model of industry-university-government partnerships throughout the analyses. Furthermore, the national economies discussed in the book (i.e., U.S., Norway, Singapore, and Qatar) represent a wide mix of economic settings and developmental levels. Similarly, the four drivers selected for the assessment of each country (i.e., legislation, institutional and cultural settings, intermediary structures, and support programs) cover all elements of knowledge ecosystem. The breadth of discussion offers the readers a broad understanding of the mechanism of these partnerships, which cover the entire spectrum of knowledge ecosystem in a country, and therefore, can benefit readers at various interest and learning levels.

Followed by the case studies, a concise comparison of the Global Innovation Index (GII) of the four countries is provided. The under-par comparative performance of Qatar is thoroughly examined through the GII indicators to learn that the country is still at the engagement level in terms of industry, university and government partnerships. To further investigate the weaknesses and potential of effective partnerships in Qatar, a comprehensive survey and interviews were carried out with the experts from academia, industry, government, and intermediaries in Qatar. Based on the results of the survey, interviews and case studies, informed policy recommendations are suggested towards the end of this book. Readers from various backgrounds, including researchers, graduate and undergraduate students, and policy makers, can benefit from the proposed evidence-based recommendations and strategies. The suggested recommendations are particularly significant for the transformation of natural capital to human capital in the developing oil and gas economies, such as Qatar.

The case of Qatar, rarely discussed before, aids in demonstrating how successful policies can potentially help an under-performing, yet committed, country in developing the basic knowledge infrastructure. The ins and outs of the knowledge ecosystem in Qatar will benefit the readers who are particularly interested in learning about the economic diversification of this country and the region in general.

Doha, Qatar   Waqas Nawaz
                                                                     Muammer Koç

Contents

Abbreviations

AAAS	American Association for the Advancement of Science
ACCM	ASTAR Collaborative Commerce Marketplace, Singapore
AHS	Academic Health System, Qatar
AI	Artificial Intelligence
AI.SG	Artificial Intelligence R&D program of Singapore
AIA	Arab Innovation Academy
AM-JIAC	Advanced Manufacturing Jobs and Innovation Accelerator Clusters
AMNPO	Advanced Manufacturing National Program Office, United States
ARC	Appalachian Regional Commission, United States
ASTAR	Agency for Science, Technology and Research, Singapore
ATIP	Agricultural Technology Innovation Partnership
ATP	Advanced Technology Program, United States
AUTM	Association of University Technology Managers
BAS	Business Angel Scheme
BIA	User-driven Resource-based Innovation
BMRC	Biomedical Research Council, Singapore
BRIO	Best Representative Image of an Outcome, Qatar
CCS	Carbon Capture and Storage
CEA	Council of Economic Advisers, United States
CEER	Centers for Environment-friendly Energy Research, Norway
CIP	Collaborative Industry Projects, Singapore
CMOS	Complementary Metal Oxide Semiconductor
CMUQ	Carnegie Mellon University in Qatar
CoE	Centers of Excellence
COI	Center of Innovation, Singapore
COT	Commercialization of Technology
CREATE	Campus for Research Excellence and Technological Enterprise, Singapore
CRI	Centers for Research-based Innovation, Norway
CWSP	Conference and Workshop Sponsorship Program, Qatar

DARPA	Defense Advanced Research Projects Agency, United States
DIC	Digital Incubation Center, Qatar
DIFI	Public Management of e-Governance, Norway
DRA	Delta Regional Authority, United States
EC	Education City
EDA	Economic Development Administration, United States
EDB	Economic Development Board, Singapore
EIA	European Innovation Academy
EQ	Enterprise Qatar
ERC	Engineering Research Centers, United States
ERIC	Energy Regional Innovation Cluster, United States
ESVF	Early Stage Venture Fund, Singapore
ETA	Employment and Training Administration, United States
EU	European Union
FCRP	Focus Center Research Program, United States
FDI	Foreign Direct Investment
FORNY	Commercializing R&D Results, Norway
FTE	Fulltime Equivalence
GCC	Gulf Cooperation Council
GDP	Gross Domestic Product
GERD	Gross Expenditure on R&D
GII	Global Innovation Index
GOALI	Grant Opportunities for Academic Liaison with Industry, United States
GPIC	Greater Philadelphia Innovation Cluster
GSA	General Services Administration, Unites States
GSRA	Graduate Sponsorship Research Award, Qatar
GU-Q	Georgetown University in Qatar
GUV	Global University Venturing
HBKU	Hamad Bin Khalifa University
HEIs	Higher Education Institutes
HMC	Hamad Medical Corporation, Qatar
HOD	Ministry of Health and Care Services, Norway
HP	Hewlett-Packard
IBCs	International Branch Campuses
IBM	International Business Machines
IC	Integrated Circuit
ICT	Information and Communication Technology
IE	International Enterprise, Singapore
IN	Innovation Norway
IoT	Internet of Things
IP	Intellectual Property
IPTT	Intellectual Property and Technology Transfer
IRD	Industrial Research and Development, Norway
IT	Information Technology

ITE	Institute of Technical Education, Singapore
iTRI	Interim Translational Research Institute, Qatar
IUCRC	Industry/University Cooperative Research Centers, United States
IUGP	Industry-University-Government Partnership
JIAC	Jobs and Innovation Accelerator Clusters
JSREP	Junior Scientist Research Experience Program, Qatar
L2 NIC	Land and Liveability National Innovation Challenge, Singapore
MCI	Ministry of Communication and Information, Singapore
MD	Doctor of Medicine
MDPS	Ministry of Development Planning and Statistics, Qatar
MEC	Ministry of Economy and Commerce, Qatar
MIT	Massachusetts Institute of Technology
MME	Ministry of Municipality and Environment, Qatar
MNCs	Multi-National Companies
MOE	Ministry of Education
MOTC	Ministry of Transportation and Communication, Qatar
MOU	Memorandum of Understanding
MRC	Medical Research Center, Qatar
MRSEC	Materials Research Science and Engineering Centers, Unites States
MSRDP	Marine Science R&D Program, Singapore
MTI	Ministry of Trade and Industry, Singapore
NASA	National Aeronautics and Space Administration, United States
NASEM	National Academy of Sciences, Engineering and Medicine, United States
NAVF	Norwegian Council for Science and Humanities
NBIA	National Business Incubation Association
NCE	Norwegian Centers of Expertise
NCR	National Cybersecurity R&D program, Singapore
NEC	National Economic Council, United States
NFFR	Norwegian Council for Fisheries
NHD	Ministry of Trade and Industry, Norway
NIC	National Innovation Challenge, Singapore
NIH	National Institutes of Health, United States
NIST	National Institute of Standards and Technology, Unites States
NLVF	Norwegian Councils for Agriculture
NNI	National Nanotechnology Initiative, Unites States
NNMI	National Network for Manufacturing Innovation, Unites States (a.k.a. Manufacturing USA)
NNN	National Nanomanufacturing Network, United States
NORAS	Norwegian Council for Applied Social Science
NPRP	National Priorities Research Program, Qatar
NPRP-EP	National Priorities Research Program (Qatar)—Exceptional Proposals
NRC	National Research Council
NRF	National Research Foundation, Singapore

NRI	Nanoelectronics Research Initiative, Unites States
NSEC	Nanoscale Science and Engineering Centers, Unites States
NSF	National Science Foundation, United States
NSRC	National Science Research Competition, Qatar
NSTB	National Science and Technology Board, Singapore
NTNF	Norwegian Council for Scientific and Industrial Research
NTNU	Norwegian University of Science and Technology
NTU	Nanyang Technological University, Singapore
NU-Q	Northwestern University in Qatar
NUS	National University of Singapore
OECD	Organization for Economic Co-operation and Development
OSTP	Office of Science and Technology Policy, United States
PACT	Partnerships for Capability Transformation, Singapore
PDRA	Postdoctoral Research Award, Qatar
PFI	Partnerships for Innovation, United States
PIC	Productivity and Innovation Credit
PPM	Path Towards Personalized Medicine Presentation, Qatar
PRICs	Public Research Institutes, Centers and Consortia
PRIs	Public Research Institutes
QAR	Qatar's Riyal
QBB	Qatar Bio Bank
QBIC	Qatar Business Incubation Center
QBRI	Qatar Biomedical Research Institute
QBWA	Qatari Business-Women Association
QCCSRC	Qatar Carbonates and Carbon Storage Research Centre
QCRI	Qatar Computing Research Institute
QDB	Qatar Development Bank
QEERI	Qatar Environmental & Energy Research Institute
QF R&D	Qatar Foundation Research and Development
QF	Qatar Foundation
QFBA	Qatar Finance and Business Academy
QIC	Qatar Innovation Community
QIPA	Qatar Innovation Promotion Award
Q-IUP	Qatar Industry-University Partnership
QMIC	Qatar Mobility Innovation Center
QNB	Qatar National Vision
QNDS	Qatar National Development Strategy
QNL	Qatar National Library
QNRF	Qatar National Research Fund
QNRS	Qatar National Research Strategy
QP	Qatar Petroleum
QRDI	Qatar Research Development Innovation
QSRTC	Qatar Shell Research and Technology Centre
QSTP	Qatar Science and Technology Park
QU	Qatar University

Q-UKRNP	Qatar-UK Research Networking Program
QUWIC	Qatar University Wireless Innovation Centre
R&D	Research and Development
R&E	Research and Experimentation
RCE	Research Centre of Excellence, Singapore
RCN	Research Council of Norway
RDA	Research and Development Tax Allowance Scheme
RDAS	Research and Development Assistance Scheme, Singapore
RIE	Research, Innovation and Enterprise
RIEC	Research, Innovation and Enterprise Council, Singapore
RISE	Research and Development Incentive Scheme for Start-up Enterprise, Singapore
ROCS	Research Outcome Center Search, Qatar
RSIM	Research Strategy and Impact Management, Qatar
RTP	Research Triangle Park, United States
SBA	Small Business Administration, United States
SBIR	Small Business Innovation Research program, United States
SEED	Start-up Enterprise Development Scheme, Singapore
SERC	Science and Engineering Research Council, Singapore
SFS	Seed Fund Support
SGD	Singapore Dollar
SI	Central Institute of Industrial Research, Norway
SIA	Semiconductor Industry Association, United States
SINTEF	Foundation for Scientific and Industrial Research, Norway
SIVA	Industrial Development Corporation of Norway
SMART	Singapore-MIT Alliance for Research and Technology, Singapore
SMEs	Small and Medium Enterprises
SMRC	Sidra Medical and Research Center, Qatar
SMU	Singapore Management University
SRC	Semiconductor Research Corporation, United States
SRING	Standards, Productivity and Innovation Board, Singapore
SSREP	Secondary School Research Experience Program, Qatar
STC	Science and Technology Centers, United States
STEAM	Science, Technology, Engineering, Art and Math
STTR	Small Business Technology Transfer program, United States
TAMUQ	Texas A&M University at Qatar
TAP	Technology Adoption Program, Singapore
TARDEQ	Tank-Automotive Research, Development and Engineering Center, United States
TECS	Technology Enterprise Commercialization Scheme, Singapore
TIP	Technology Innovation Program, United States
TTOs	Technology Transfer Offices
T-Up	Technology for Enterprise Capability Upgrading, Singapore
UCL-Qatar	University College London, Qatar
UNESCO	United Nations Educational, Scientific and Cultural Organization

UREP	Undergraduate Research Experience Program, Qatar
US DA	Department of Agriculture, United States
US DOC	Department of Commerce, United States
US DOD	Department of Defense, United States
US DOE	Department of Energy, United States
US DOL	Departments of Labor, United States
US ED	Departments of Education, United States
USA/US	United States of America
USD	United States Dollar
VCUarts	Virginia Commonwealth University School of the Arts in Qatar
VRI	Regional R&D and Innovation, Norway
WCM-Q	Weill Cornell Medical College in Qatar

Chapter 1
Introduction to Industry, University, and Government Partnerships: Theoretical Model

Abstract The world has recognized the role and significance of innovation for the economic and social development and national prosperity. The primary players involved in synergizing the innovative capabilities and outcomes for establishing a knowledge-based ecosystem are industries, universities, and government. The partnerships between industry, university and government create opportunities to translate the fundamental research into value-driven products and services. The mechanism of these partnerships has evolved over time—from statist and Laissez faire to the modern triple helix model, where the roles of industries, universities and government are balanced but interdependent and supportive of each other. The key drivers of knowledge-intensive development include institutional and cultural settings, legislations or regulations, support programs, and the promotional structures and mechanisms. Besides introducing the triple helix model and the drivers of knowledge ecosystem, this chapter will present the objectives, motivations, and the organization of chapters in this book.

1.1 Industry, University, and Government Partnerships (IUGP) as a Framework for Innovation Capacity Building

Traditionally, innovation has been modeled as a process that undergoes a linear sequence of research, development, production and marketing. However, in recent years, innovation has been conceptualized as a result of complex interactions between various actors of economic and social processes such as the enterprises, academic institutions, laboratories, and the consumers, as well as a feedback between science, engineering, product development, manufacturing and marketing (Organization for Economic Cooperation and Development OECD 1996; Manley 2002). Innovation has gained an increased importance in advanced and emerging economies due to the knowledge intensive nature of the growth processes.

Education in knowledge economy has key functions of (knowledge) production, transmission and transfer, and contributes substantially to the competitiveness of

© Springer Nature Switzerland AG 2020

W. Nawaz and M. Koç, *Industry, University and Government Partnerships for the Sustainable Development of Knowledge-Based Society*, Management and Industrial Engineering, https://doi.org/10.1007/978-3-030-26799-5_1

industries (Mansfield 1995; Mansfield and Lee 1996). In addition to the two basic missions of teaching and research, entrepreneurial universities engage in the third mission (Larédo 2007; Tijssen 2006) of contributing to society by diffusing knowledge to the local community, thereby generating research and consultancy income and assisting domestic and regional development. With this new mission, support through government policies and incentives have also become instrumental, and hence Industry-University-Government Partnership (IUGP) has been conceptualized and developed in various countries based on their cultural and political and innovation frameworks. Nevertheless, it is also crucial to recognize that the success of third mission's activities depend on both the specialization of higher education organizations and the characteristics of the local context (Sánchez-Barrioluengo 2014). Indeed, indicators of the innovative capacity of regions in the European Union (EU) show a strong correlation between highly skilled human resources, technological infrastructure, and creativity. A recent study argues that "knowledge created in universities and other institutes of higher education is essential for the development of required skills and expertise; it also allows companies to have access to state-of-the-art laboratories and capable graduates" (Sleuwaegen and Boiardi 2014). Consequently, there is a growing interaction between the educational system and the private sector in conducting basic and applied research with the support of government agencies.

International experience suggests that IUGPs can play a proactive role in technology transfer, especially by engaging universities more closely with industry associations, leaders, and also the Small and Medium Enterprises (SMEs)—the most challenging category of industry players. Structural characteristics of the industries and universities such as size of Research and Development (R&D), structure/discipline, technology absorption/transfer capacity and market/organization structure represent the demand and supply of the national knowledge market. The rationality between the demand and supply determines the potential need of the IUGP and shapes the incentives and barriers for the relevant actors (Polt et al. 2001). Past experiences show that IUGPs are specific to certain economic sectors and fields of technology. Most literature (Mowery and Sampat 2005; Perkmann et al. 2013) identify that the effect of these partnerships is prominent in case of young technological fields such as biotechnology and Information Technology (IT) in comparison to the well-established fields with high market competition. To a considerable degree, the effectiveness of IUGPs depends on the organizational set-up and responsiveness with regards to the industry partners (Siegel et al. 2003). Carlsson and Fridh (2002) argue that the success of a technology transfer process depends not only on the interface between the university and the business community but also on the receptivity in the surrounding community and the culture.

1.2 Triple Helix Model

The recent IUGP literature provides a set of theoretical and methodological approaches for understanding the role of universities and research institutions in advanced economies, and the influence of policies designed to enhance the commercialization of technology through university-industry linkages. Efforts to develop a better understanding of the role of universities and research institutions has increasingly been conceptualized within the triple helix model of innovation-driven economic growth (Etzkowitz 2008; Ranga and Etzkowitz 2013). The present interactive configuration of triple helix model is a result of continuous transformation of the model through constant learning and improvement. Figure 1.1 shows three prevalent configurations of the IUGP model.

The statist configuration is a classic configuration in which government plays the lead role (Etzkowitz 2002; Ranga and Etzkowitz 2013). IUGP structures in Russia, China and some Latin American and Eastern European countries previously had such configurations. The configuration turned outdated with the realization of its potential to limit the capacity of development and innovative transformation in a country. On the other hand, Laissez faire configuration is characterized by limited roles of government and academia in innovation; industry acts as a driving force and universities and government provides the necessary supports. United States and some Western European countries had such IUGP configuration previously (Ranga and Etzkowitz 2013).

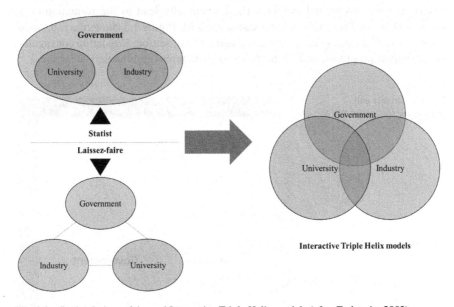

Fig. 1.1 Statist, Laissez-faire and Interactive Triple Helix models (after Etzkowitz 2002)

In both statist and Laissez faire configurations, the main role of university is to create high skilled human capital and to catalyze new knowledge, while industry works for stimulating knowledge and innovation, and government plays as a facilitator in knowledge creation and diffusion and innovation through various policies and implementation instruments (Etzkowitz and Leydesdorff 2000). However, in today's world, the roles of the three actors have been redefined in a more interrelated way. For example, universities now a days are known for their start-ups and spin-offs, industries provide education through trainings and internships, and government acts as a venture capitalist (Etzkowitz et al. 2000). This transition from the traditional model has led to an interactive configuration of the triple helix model—characterized with balanced but interdependent roles of industry, university and government.

1.3 Drivers of Industry, University, and Government Partnerships

Advanced and emerging economies have adopted different initiatives to strengthen interactions between industry, university and government to promote innovative competitiveness. Regional clustering and strategic programs such as workforce training, entrepreneurship programs, and technical and financial assistance have been widely adopted. These initiatives give rise to different types of IUGPs such as collaborative research, co-operation for innovation, start-ups for innovative ideas, flow of researchers and vocational training. These eventually lead to the formation of a framework for building national innovation capacity. Figure 1.2 shows the four main drivers of IUGP, namely the institutional and cultural setting, legislations or regulations, support programs, and the promotional structures (intermediary structures).

Fig. 1.2 Drivers and settings for successful IUGP

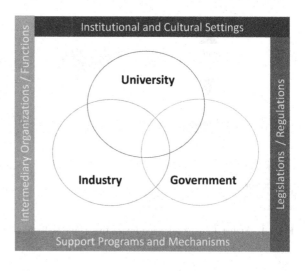

IUGPs show different mechanisms under different institutional and cultural settings; a setting is the result of historical developments in social, economic, and political aspects of the institution. Countries around the world have practiced some different, but many common, regulatory and policy initiatives. Policy initiatives to promote IUGPs have two main objectives. Firstly, it aims to ensure that the public investments in universities/research institutions spill over to the industries and make economically productive results. Secondly, it is expected to act as a tool for industries to raise their competitiveness and technological performance by effectively absorbing the innovative knowledge.

Most countries launch their policy initiatives in the form of intermediary structures or support programs. Intermediary structures cover a wide range of infrastructures established to foster industry-university partnerships. Broadly, the structures are managed by specialized professionals, created to stimulate the regional economic development and to support the commercialization of technology inventions (OECD 2002). These structures improve the return on R&D investments through active collaboration among innovation actors and have the potential to provide global leadership in key technologies. Some of the examples include government/semi-government funded R&D centers/institutes/clusters, science/technology parks, incubators, and technology transfer offices.

For most countries, research institutes and innovation centers are the result of the early government support to enhance business innovation. More recently, clustering or grouping of enterprises and related research communities for a common vision of development has become increasingly popular among governments, academia and industries. In this regard, regional clustering of science and technology centers, research institutes, and business parks are gaining popularity among emerging economies.

Technology Transfer Offices (TTOs) are specialist entities in universities, research institutions, governments or companies that facilitate commercialization of the fundamental research through licensing the intellectual property to SMEs or start-up companies. Business incubation centers, also known as business incubators, are entities that help viable start-ups and early-stage companies to become self-sustaining by providing office space, managerial training, and financial and technical support.

Governments in many countries use their purchasing power to provide various financial incentives to leverage demand for domestic industries. Another widely used incentive is the tax credit or deduction in tax to promote emerging technologies. Universities offer internship, entrepreneurship and training programs, and innovation-oriented curriculum to support national human development strategies. Additionally, several focused support programs are established by the governments/universities/industries to facilitate business innovation through public-private partnerships (PPPs). Interaction between public research and industry is crucial to fill the gap between high-quality scientific research and the application of the research outcomes. One of the advantages of the partnership programs is that it is readily accessible by academia and industries and is appropriate for a wide range of industry-university partnerships compared to the intermediaries which require ambitious initial effort to create a center that needs multiple industrial partners and faculty research teams to support it.

1.4   Objective and Motivation

With the emerging concerns and vision for the sustainable development of Qatar, a country that combines interdependent issues of economy, energy, environment and human development, it has become apparent that the solution intrinsically lies in the creation and dissemination of knowledge and innovation. The need for innovation affects governments, businesses, and individual citizens in ways that may demand new expenditures but that can also offer new opportunities for entrepreneurship, especially green entrepreneurship. This situation is a challenge for Qatar as much as it is for other societies. The Government of Qatar has recognized the challenge in its National Vision (QNV) 2030, emphasizing all pillars of sustainable development, i.e., human, social, environmental and economic development, in a balanced manner.

For an emerging economy such as Qatar, the need to promote closer collaboration between industry and universities not only represents a major policy and cultural challenge, it is a unique opportunity to examine theoretical ideas about innovative engagement in a society that is rapidly moving, or at least aims to move, towards the knowledge economy.

Abduljawad (2015) and Mohtar (2015) have shed some light on the industry-university partnerships in Qatar, however, their work was limited to a single sector. In this book, we provide a holistic account on the IUGP trends, drivers and incentives in Qatar in comparison to the theories, models and best practices from other parts of the world. In this book, we offer evidence-based recommendations for the improvements of IUGPs in Qatar. The study has three basic components: (i) comparison of IUGP trends, drivers and incentives in Qatar with other countries, such as US, Norway, and Singapore; (ii) comparison of Qatar with other countries over the global innovation indicators, and (iii) views and expectations of professionals with first-hand experience of working in the IUGPs in Qatar.

1.5   Organization of Book Chapters

Chapters 2, 3 and 4 explore the IUGP trends and drivers in three advanced and emerging economies: United States of America, Norway, and Singapore, respectively. Chapter 5 provides an overview of the IUGP structure in Qatar. Chapter 6 provides a comparative account on the IUGP settings of the four countries. It also offers a comparison among the four countries on various Global Innovation Indicators. Chapter 7 covers the results of the survey and interviews conducted with experts in Qatar to further explore the innovation trends, challenges, and improvement opportunities for the country. Chapter 7 also provides a list of policy recommendations to improve the effectiveness of IUGPs in Qatar.

References

Abduljawad, H. (2015). Challenges in cultivating knowledge in university-industry-government partnerships—Qatar as a case study. *The Muslim World, 105*(1), 58–77.

Carlsson, B., & Fridh, A. C. (2002). Technology transfer in United States universities—A survey and statistical analysis. *Journal of Evolutionary Economics, 12*(1–2), 199–232.

Etzkowitz, H. (2002). *The triple helix of university-industry-government: Implications for policy and evaluation.* Retrieved from https://scholar.google.com/scholar?q=The+Triple+Helix+of+University+-+Industry+-+Government+Implications+for+Policy+and+Evaluation&btnG=&hl=en&as_sdt=0%2C5.

Etzkowitz, H. (2008). *The Triple Helix: University-Industry-Government innovation in action.* London: Routledge.

Etzkowitz, H., Gulbrandsen, M., & Levitt, J. (2000). *Public venture capital: Government funding sources for technology entrepreneurs.* New York: Harcourt-Brace.

Etzkowitz, H., & Leydesdorff, L. (2000). The dynamics of innovation: From National Systems and "Mode 2" to a Triple Helix of university–industry–government relations. *Research Policy, 29*(2), 109–123.

Larédo, P. (2007). Revisiting the third mission of universities: Toward a renewed categorization of university activities? *Higher Education Policy, 20,* 441–456.

Manley, K. (2002). The systems approach to innovation studies. *AJIS, 9*(2), 95–102.

Mansfield, E. (1995). Academic research underlying industrial innovations: Sources, characteristics, and financing. *Review of Economics and Statistics,* 55–65.

Mansfield, E., & Lee, J.-Y. (1996). The modern university: Contributor to industrial innovation and recipient of industrial R&D support. *Research Policy, 25,* 1047–1058.

Mohtar, R. H. (2015). Opportunities and challenges for innovations in Qatar. *The Muslim World, 105*(1), 46–57.

Mowery, D. C., & Sampat, B. N. (2005). The Bayh-Dole Act of 1980 and university–industry technology transfer: A model for other OECD governments? *Journal of Technology Transfer, 30*(1/2), 115–127.

OECD. (1996). *The knowledge-based economy.* Paris: OECD.

OECD. (2002). *Benchmarking science-industry relationships.* Paris: OECD.

Perkmann, M., Tartari, V., McKelvey, M., Autio, E., et al. (2013). Academic engagement and commercialisation: A review of the literature on university-industry relations. *Research Policy, 42,* 423–442.

Polt, W., Gassler, H., Schibany, A., Rammer, C., & Schartinger, D. (2001). Benchmarking industry-science relations: The role of framework conditions. *Science Public Policy, 28*(4), 247–258.

Ranga, M., & Etzkowitz, H. (2013). Triple Helix systems: An analytical framework for innovation policy and practice in the knowledge society. *Industry & Higher Education, 27*(3), 237–262.

Sánchez-Barrioluengo, M. (2014). Articulating the 'three-missions' in Spanish universities. *Research Policy, 43,* 1760–1773.

Siegel, D. L., Waldman, D., & Link, A. (2003). Assessing the impact of organizational practices on the relative productivity of university technology transfer offices: An exploratory study. *Research Policy, 32,* 27–48.

Sleuwaegen, L., & Boiardi, P. (2014). Creativity and regional innovation: Evidence from EU regions. *Research Policy, 43,* 1508–1522.

Tijssen, R. J. W. (2006). Universities and industrially relevant science: Towards measurement models and indicators of entrepreneurial orientation. *Research Policy, 35,* 1569–1585.

Chapter 2
Case Study: United States of America

Abstract United States of America is of a special interest in academic studies due to its size, population, resources, and the development quotient. The country has organically advanced in the paradigm of research and innovation through advanced, diverse and strong IUGPs, which resulted in the creation of knowledge-intensive business opportunities and jobs. This chapter provides a comprehensive account on the development of IUGPs in the US. First, we explore the history of the IUGPs in the US—how it evolved and who supported it? Second, we discuss the legislation around the IUGPs, such as the Bayh-Dole Act which is one of the widely credited acts for improving university-industry collaboration and technology transfer in the US national innovation system. Third, we take account of the intermediary structures in the US which support the translation of research results into commercialized products/services, such as the Industry-University Cooperative Research Centers (IUCRCs), Engineering Research Centers (ERC), research parks, and industrial innovation centers. Finally, we review the national policies that encourage the collaboration between universities, industries, and government, such as the public procurement of integrated circuit chips, research and experimentation tax credit program, and small business innovation research program.

2.1 Background

United States' economy is one of the highly developed economies in the world; its nominal Gross Domestic Product (GDP) and GDP per capita (in Purchasing Power Parity) in 2016 were 18.62 trillion USD (highest in the world) and 57,638 USD (thirteenth highest in the world), respectively (The World Bank 2017). It is also one of the most populated countries in the world; as of July 2016, it had a population of more than 323 million (US Census Bureau 2017), and was ranked as the country with third highest population in the world (The World Bank 2017). The US economy is strongly backed by its abundant natural resources and high productivity supported with highly advanced technology and infrastructure. R&D has been a core sector of

---

R. Pradhananga co-authored this chapter.

© Springer Nature Switzerland AG 2020

W. Nawaz and M. Koç, *Industry, University and Government Partnerships*
for the Sustainable Development of Knowledge-Based Society,
Management and Industrial Engineering, https://doi.org/10.1007/978-3-030-26799-5_2

importance from its early years of development. As of 2013, the R&D expenditure of the country is 2.73% of its GDP and more than 60% of this was funded by the business sector (Eurostat 2017). A large share of federal R&D funding goes to universities, which has a significant role in the technological advancement of the country. For 2017, the Global Innovation Index (GII) of US was 61.40 which was the fourth highest among the 127 countries across the world (Cornell University, INSEAD, and WIPO 2017). Strong academic circles and their strong linkage to industries, backed by effective government support, are major contributors to the success of the country.

2.2 Institutional and Cultural Setting

Prior to World War II, US innovation process was decentralized and highly market-driven. Private sectors and charitable foundations were the main contributors and federal government had limited role. Research activities were mainly concentrated to few elite universities and few private firms. Most universities faced lack of centralized administrative control, considerable inter-institutional competition and reliance on state level sources for political and financial support. The potential political and financial support motivated the universities to collaborate with the regional industries, and the lack of central control forced universities to be more entrepreneurial (Ben-David 1968; Goldfarb et al. 2001), initiating a trend of regional clustering often named as science/technology/industry parks in the US national innovation system.

The success of university-initiated clusters in the US motivated other advanced and emerging economies to investment in R&D and to implement supportive policies to encourage cluster development. In US, regional clustering was soon significantly supported by the state and local governments. Governments in many states such as North Carolina, New York, South Carolina, Ohio, New Mexico, and Michigan began to develop comprehensive cluster-based strategies to create new sources of high-wage jobs (National Research Council, US NRC 2011) and over the time these initiatives helped the states to become one of the world's premier hubs of R&D. A wider range of intermediaries including technology transfer offices and business incubation centers were established. Policy tools such as tax credits, R&D grants, risk capital to start-ups and free or subsidized workforce training were deployed to provide financial incentives to the industries and research institutions. Despite these growing efforts, federal participation in regional clustering during this period was limited. The strong intellectual property rights, risk taking financial incentives, flexible labor force, and openness to foreigners and an entrepreneurial culture kept US in a leading position in innovation (US NRC 2012).

Federal government support in innovation and technology transfer primarily comes in the form of R&D investments to the universities and the national laboratories. R&D funding from the federal government is coordinated by several regulatory and non-regulatory federal agencies such as the Departments of Education (US ED), Defense (US DOD), Energy (US DOE), Commerce (US DOC), Agriculture (US

DA), the National Institutes of Health (NIH), the National Aeronautics and Space Administration (NASA), National Institute of Standards and Technology (NIST), and the National Science Foundation (NSF). These agencies allocate research grants for basic and applied researches on a peer-review basis.

Federal government has a long history of supporting public-private partnership by sponsoring various strategic programs. Small Business Innovation Research (SBIR) Program is an exemplary initiative that support competitive small businesses providing start-up funds. NSF is sponsoring dedicated industry-university cooperative program, Industry-University Cooperation Research Centers (IUCRC), since 1970. Also, several state governments have invested in public-private research institutions to stimulate local manufacturing industries. More recently, federal government is initiating several strategic programs to support national level clustering. US DOD, US DOE, US DOC, US DA together with Departments of Labor (US DOL) and US ED have launched and funded various programs to support creation of new clusters and coordination of the existing state and regional cluster initiatives.

The US ED is supporting this innovative trend for long-term sustainability and has implemented several grant programs to serve innovative and resilient culture in its higher education system. Education Innovation and Research program, Investing in Innovation, Ready to Learn Television, and Skills for Success are the four programs in this direction, operational under Office of Education Innovation Programs (US ED OII 2017).

Over the time, US national innovation system has evolved as a robust and advanced system comprising of an institutional setting with diverse forms of partnerships between the government, academia, and industries. The US congress has power to introduce innovation related legislations. The federal agencies operating through the federal system have power to create, fund, and coordinate policy programs for the innovation-related missions. The federal government shares such power with state and local governments as well. Innovations are disseminated to private sector through multiple methods, including supply chain, licensing of Intellectual Property (IP), and movement of human capital (Shapira and Youtie 2010). Intermediary structures play important role in facilitating learning and transferal of innovation practices, taking on technology transfer roles and becoming hubs for incubators, spin-offs and knowledge transfer.

2.3 Legislations/Regulations

Much of the landmark legislations relating to US innovation came after 1980. Shapira and Youtie (2010) provides a list of such legislations in chronological order. Bayh-Dole Act is one of the widely credited acts for improving university-industry collaboration and technology transfer in the US national innovation system. Before the Bayh–Dole Act, research projects with federal funding obligated inventors to assign inventions to the federal government. Enactment of Bayh-Dole Act in 1980 facilitated university patenting and licensing (Carlsson and Fridh 2002; Chai and Shih

2016), and permitted university, small business, and non-profit institutions to elect for the ownership of an invention in preference to the government (Blackwell 2012). It brought a broader shift in US policies toward stronger IP rights. University patenting in US increased significantly after the enactment of the Bayh-Dole Act. In contrast, some argue that much of the growth in university-based licensing and patenting would have occurred even in the absence of the Bayh-Dole Act, as US universities were actively involved in patenting and licensing of innovations decades before 1980 (Mowery and Sampat 2005). Many countries considered policy initiatives that emulate the Bayh-Dole Act. The international efforts to emulsify Bayh-Dole Act, however, suffered from various cultural barriers. According to Mowery and Sampat (2005), enhancement of inter-institutional competition, autonomy of universities, promotion of new firms and technology commercialization can accelerate innovation more effectively over such emulations.

In addition to the Bayh-Dole Act, Stevenson-Wydler Technology Act of 1980 made it compulsory for the federal laboratories to establish and fund TTOs to facilitate transfer technology to nonfederal entities, and to provide means to other organizations (outside government bodies) for accessing federal laboratory technologies. Small Business Innovation Development Act of 1982 requires federal agencies to provide 2.5% of its extramural budget for domestic small business R&D that has the potential for commercialization. The act created Small Business Innovation Research (SBIR) program, one of the best practices in public-private partnership (US, NRC 2013), and was reauthorized in 2000 and 2008. National Cooperative Research Act of 1984 and Federal Technology Transfer Act of 1986 encouraged joint ventures in R&D among industries, universities and federal laboratories. Omnibus Trade and Competitiveness Act of 1988 strengthened the competitiveness of US companies through changes in process of trade law and gave a new role to US DOC in technology transfer and innovation. These legislations developed in 1980s facilitated intellectual property protection, technology transfer, small business innovation and joint ventures. Legislations developed in later periods expanded these legislations. For example, the American Technology Preeminence Act of 1991 and the National Technology Transfer Improvements Act of 1995 extended the intellectual property protection. The Small Business Technology Transfer Act of 1992 established the Small Business Technology Transfer program to facilitate commercialization of universities and federal laboratories inventions through small businesses.

Among the legislations passed in the last two decades, America COMPETE Act of 2007 is one of the key initiatives towards coordination of innovative efforts from different federal agencies to ensure an effective system-wide governance. The aim was to strengthen framework conditions for business innovation providing research investments, opening educational opportunities in science and technology and supporting greater infrastructure for innovation management (Shapira and Youtie 2010). Reauthorization of the act in 2010 established national and regional innovation programs to support innovation strategies. The reauthorized act increased the federal research budget and the coordination efforts. The act funds high risk and high reward projects and multi-agency collaborations to support clustering efforts. Another prominent legislation is the Leahy-Smith America Invents Act of 2011, which brought major changes in the US patent system. The act shifts US patent system's emphasis from "first to invent" to "first to file" system. The new system places an emphasis on the date of the application of the patent rather than on the date of invention.

2.4 Intermediary Structures

Following describes the various categories of intermediary structures established in US to strengthen research and innovation.

2.4.1 Government/Semi-government Research Institutions

Throughout the history, federal agencies have facilitated creation of university/industry-led centers/hubs/institutes of innovations in different disciplines supporting joint efforts of academia, industry and government. Some examples include Industry/University Cooperative Research Centers (IUCRC), Science and Technology Centers (STC), Materials Research Science and Engineering Centers (MRSEC), Engineering Research Centers (ERC), Nanoscale Science and Engineering Centers (NSEC), Nanoelectronics Research Initiative (NRI) Research centers, the i6 challenge: Proof of Concept Centers, Focus Center Research Program (FCRP) Centers, and the National Network for Manufacturing Innovation (NNMI) Institutes. The examples will be discussed in the following subsections.

2.4.1.1 IUCRC

IUCRC is NSF sponsored program dedicated to industry-university cooperation and is in place since 1970s (NSF IUCRC NIST 2017). Currently, there are more than 150 IUCRCs in more than 100 universities focusing on applied research issues including advanced electronics, photonics, advanced manufacturing, advanced materials, biotechnology, civil infrastructure systems, energy and environment, health and safety, information, communication and computing, and system design and simulation. These IUCRCs conduct the industry driven research projects in support from more than 1500 companies including both large, Multi-National Companies (MNCs), as well as domestic small businesses coming out of university faculty and students and/or local entrepreneurs. Each IUCRC conducts several R&D projects on a non-confidential, shared and royalty-free basis for their industry members in return for an annual fee varying according to the size of company members. Universities support such centers by providing additional lab-space, equipment funding, student tuition waiver or reductions, overhead waiver or reductions, and most importantly recognition. Government support comes mainly through NSF by putting a seed fund at the very beginning of the formation of an IUCRC, which is usually started by faculty (director), and annual funds up to ten years and a five-year extension can be awarded depending on the industry interest and membership. According to the recent surveys conducted independently by the department of psychology of the North Carolina State University (NCSU), IUCRCs' intellectual efficiency is highest in terms of the patents or publications per faculty and/or per dollar (NCSU 2017). Another major advantage of IUCRC is the direct impact of research and the use of research results to support the end users (industry).

2.4.1.2   ERC

ERCs are multi-institutional centers created under the NSF funded programs. Established in 1985 (NSF ERC 2015), ERC programs focus on research, education and industrial interaction with an objective to facilitate creation of next-generation industries through radical advancement of the current engineering systems along with production of a new generation of engineering graduates proficient to lead these industries (NSF ERC 2017). Each center is jointly formed between universities (the host of the center), industries and government. NSF funds the ERCs for a maximum of ten years. After the support and funding period, ERCs are expected to become self-sustaining entities. Currently, there are 19 active ERCs covering four broad technology areas: (i) advanced manufacturing; (ii) microelectronics, sensing and information technology; (iii) energy, sustainability, and infrastructure; and (iv) biotechnology and health care. Lewis (2010) evaluated the economic impacts of ERC-generated technologies on the US economy. The author reported the market value of benefits created through ERCs as tens of billions of dollars. In addition, engagement with ERCs increased employability of the students and graduates.

2.4.1.3   STC

Established since 1987, STC integrative partnership program is also funded by the NSF. The aim of STC is to support partnerships among universities, industries, national laboratories, and/or other public/private entities to conduct innovative and potentially transformative research and education projects that require large and long-term financial support (NSF STC 2017). The program has supported 51 STCs from a wide range of areas of science and technology. Currently, 12 of the STCs are active. The host of the institution, which is one of the partnering institutions, is responsible for the management and finances of the center. NSF funds STC for a maximum of ten years; five years as initial commitment and possible continuation for another five years (NSF STC 2014). Such long-term financial support from NSF provides reasonable opportunity to STCs to achieve their scientific objectives (US NRC 1987). American Association for the Advancement of Science (AAAS) reported outstanding performance of the centers in terms of research transformation, collaborations, expansion of science and technology workforce, and most importantly, in bringing workforce diversity (Chubin et al. 2010).

2.4.1.4 MRSEC

MRSECs, established in 1994, consists of network of university-based centers in US funded by the NSF (US NRC 2005). The centers promote active collaborations between universities, industries, and other national and international institutions to conduct research and education projects of intellectual and technological significance in the area of material science and engineering (NSF MRSEC 2017). Apart from conducting high quality research, the centers develop shared experimental and computational facilities, involve in educational and outreach activities, provide seed funding for transfer of results to industries, and collaborate with a diverse group of practitioners in the field. NSF funds MRSECs for a period of six years depending on its periodic progress (NSF MRSEC NSF 2016). Currently there are 21 MRSECs across the US. The MRSEC can be a small center with a focused topic or a larger center with broader program depending on the capabilities of the base university and the partnering institutions. The MRSEC impact assessment results published in 2005 reported that the MRSECs have been successful in making an impact of the same (high) standard as the other NSF supported programs (US NRC 2005).

2.4.1.5 FCRP Center

FCRP was a government-industry co-sponsorship program launched in 1997 to strengthen microelectronics research capability of US universities (Semiconductor Industry Association, SIA 2010). In 2013, the activities of the program were discontinued under FCRP (in its fifth cycle); the activities were transferred to the STARnet program, also referred as FCRP VI, with an aim to sponsor long-term research projects in the area (Semiconductor Research Corporation, SRC FCRP 2017). From its inception to the end, the program established six FCRP centers, each one dedicated to one of the major technology focus areas of the International Technology Roadmap for Semiconductors. Each center consists of multiple universities that conduct exploratory research matching the center's objective. The centers are characterized by university-driven management philosophy, substantial funding to encourage non-traditional approaches, student training, and sustained industry commitment (US NRC 2003). Semiconductor Research Corporation (SRC) and Defense Advanced Research Projects Agency (DARPA) administered FCRP centers, and the US semiconductor and supplier industries and US DOD jointly funds the research projects. The FCRP centers benefit the member companies to be more competitive providing access to the technological discoveries in the universities. As of 2017, 49 universities participated in the FCRP program and published 11,402 papers and filed 354 patent applications, of which 130 have been granted (SRC FCRP 2017).

2.4.1.6 NSEC

NSEC program is also an NSF initiative to complement the National Nanotechnology Initiative (NNI), which is a federal program dedicated to the R&D in nanoscale science, engineering and technology (Batterson 2002). In 2001, the first group of six NSECs were created that were led by six outstanding research universities in the US. According to National Nanomanufacturing Network (NNN), currently there are 18 NSF funded NSECs across the US focusing nanoscience researches in multidiscipline areas of material science, chemistry and biomedical sciences (NNN 2017). NSF provides fund to the centers for five years with possible extension of the funding for additional five years as renewal (NSF NSEC 2004). The centers are led by US academic institutions with undergraduate and Ph.D. programs. At NSECs, researchers with diverse expertise collaborate with industries, government laboratories and other public and private sector organizations to conduct complex multi-faceted projects in nanoscale science and engineering. The centers focus on education, human capital development and outreach programs, and provide seed funding for commercialization of high-risk research. Assessment of the NSECs by Rogers et al. (2011) documented outstanding scientific performance of the centers—evident from the publications of results in high impact journals. Furthermore, some centers developed significant volume of commercial activities as a result of partnerships with organizations in various other sectors.

2.4.1.7 NRI Research Center

NRI was launched by the Semiconductor Industry Association (SIA) in 2005 with an aim to develop electronic components that can replace conventional CMOS (Complementary Metal Oxide Semiconductor) and extend today's computer technology (NIST NRI 2012). Under this program, four university-led research centers were established which received direct support from NIST for metrology expertise (SIA 2012). NIST and the state governments offer additional funding support to these centers, if required. Each of the four centers have shown significant patent activities and published several scientific papers in high impact journals (NRI 2012).

2.4.1.8 The i6 Challenge: Proof of Concept Centers

The i6 challenge is a multi-agency federal program led by US DOC's Economic Development Administration (EDA). Launched in 2010, the program facilitates innovation, technology commercialization, entrepreneurship, and partnerships to support the regional economies (EDA The i6 Challenge 2012; Bradley et al. 2013). Under the program, EDA provides funding to universities and research centers to create proof of concept centers that help commercialization of the innovations from universities and promote partnerships for new enterprise formation. Each center receives guaranteed funding for a period of two years which can be extended further by the

EDA depending on the circumstances. Also, the centers get significant supports from other federal agencies such as US DOE, US DA, NIH, Small Business Administration (SBA), NIST, NSF, and US Patent and Trademark Office (US PTO). EDA approved 12 proof of concept centers through two rounds of funding in 2010 and 2011, covering a wide range of areas such as medical technology, bioscience, nanoscience, drug development, renewable energy, and green building technology. SRI International, an independent nonprofit research center, assessed the performance of these centers by conducting a survey with businesses and organizations which received support or services from the proof of concept centers (SRI International 2014). More than 60% of the respondents reported improvements in technology transfer and commercialization activities due to the support from the proof of concept centers.

2.4.1.9 NNMI Institutes

NNMI, also known as Manufacturing USA, is a federal program established in 2014 with an aim to bring together industry, academia, and federal partners to enhance US industrial competitiveness and economic growth with cutting-edge manufacturing sector (Manufacturing USA 2017). The program is operated by Advanced Manufacturing National Program Office (AMNPO) and is hosted by NIST in partnership with US DOD, US DOE, NASA, NSF, US ED and US DA. As of May 2019, NNMI has 14 university/industry-led institutes. While all these institutes have a common goal to link innovation and manufacturing, each institute focus on a unique area of R&D. Each institute facilitate knowledge transfer in scale-up activities, from laboratory to manufacturing phase. The scale-up activities include applied research, technology development, prototyping, education and workforce development, and outreach programs for SMEs and large-sized manufacturing firms (NNMI 2016). NNMI institutes also receive support from state and local governments in establishing projects with direct benefits to the communities. NNMI institutes are funded by the federal and non-federal agencies in a cooperative agreement for five years. The institutes are expected to become self-sustaining following the funding period. Assessment of the institutes at current stage would be inappropriate as most of the institutes are newly established. However, after just few years of establishment, many of the institutes have already organized far-reaching consortiums and supported several excellent technology and workforce development programs.

2.4.2 Government Initiated Innovation Clusters

The idea of regional clustering or grouping gained rapid popularity among US state governments in the late 1950s. However, many states attempted to launch clusters in the same industries, such as biotechnology—due to the lack of coordination (US NRC 2012). The redundancy of the clusters identified a strong need for the national programs to support and coordinate the existing regional clustering efforts. Federal programs to support clustering started remarkably after the enactment of reauthorization America COMPETES Act in 2010. Under these programs, federal agencies, such as US DOE, US DOC, US DOD, US DA, US DOL, US ED, SBA and NSF, independently, and in collaboration, lead the creation and coordination of innovation clusters in diverse high technology sectors. Some of these clusters include the Agricultural Technology Innovation Partnership (ATIP) cluster, Energy Regional Innovation Cluster (ERIC), SBA's Pilot Contract based Cluster, Jobs and Innovation Accelerator Clusters (JIAC), Advanced Manufacturing JIAC (AM-JIAC), and rural JIAC.

2.4.2.1 ATIP Cluster

Established in 2007 by US DA, the ATIP program is dedicated to enhancing the commercialization potential of US DA's research outcomes (US DA 2011). Currently, ATIP consist of a cluster of ten members; a team of nine partners and one associate. The cluster helps US DA to enhance its technology transfer ability through the skills, knowledge and capabilities of ATIP members (US DA 2011). ATIP is connected with the universities and small business development centers to strengthen its partnerships. The cluster helps the private sector in introducing new products and technologies, conducting market research, and facilitating licensing applications and commercialization. National Economic Council (NEC), Council of Economic Advisers (CEA), and Office of Science and Technology Policy (OSTP) documented ATIP as an example cluster initiative of US DA to enhance regional innovation in agriculture technology sector (NEC, CEA, OSTP 2011).

2.4.2.2 ERIC

In 2010, US DOE started ERIC program to create regional innovation clusters in solar power, batteries, nuclear energy and energy-efficient buildings (Johnson 2012). The clusters are expected to work in close coordination with DOE's energy innovation hubs, coordinate and accelerate other regional innovation programs and initiatives, and stimulate private investment and quality job creation. Greater Philadelphia Innovation Cluster (GPIC) is the first ERIC for energy efficient buildings. DOE is leading six other federal agencies, including SBA, DOC's NIST, US ED, EDA, NSF and US DOL, to coordinate this cluster (US DOE 2010).

2.4.2.3 SBA's Pilot Contract-Based Cluster

In 2010, SBA launched its pilot contract-based cluster program with a mission to link small businesses to regional networks of leading research, commercialization and financing in a most effective way. Under the program SBA funded ten innovation clusters, seven of which are regional innovation clusters from a wide range of geography and leading technologies, and rest of the 3 are advanced defense technology clusters in the areas of critical importance to US DOD (SBA 2017a). These clusters support small businesses with mentoring, counseling, and guiding mechanisms of technology transfer and commercialization. In 2014, SBA announced funding of four additional regional clusters in the areas of water technology, marine industries science and technology, autonomous and unmanned systems, and retail, supply chain and food processing (SBA 2014a). Evaluation of SBA's pilot contract-based clusters, performed by Optimal Solutions Group, shows a remarkable increase in the small business participation with the cluster initiative (Optimal 2012).

2.4.2.4 JIAC, AM-JIAC, Rural JIAC

With a mission to accelerate job creation and develop skilled workforce, the federal government has launched three multiagency cluster development programs, JIAC, AM-JIAC and the rural JIAC (EDA 2016a). In 2011, US DOL's Employment and Training Administration (ETA), EDA, and SBA funded 20 industry clusters under the JIAC program. Furthermore, in 2012, EDA, ETA, SBA, NIST and DOE funded ten advance manufacturing industry clusters under the AM-JIAC program. In the same year, EDA, US DA, Delta Regional Authority (DRA) and Appalachian Regional Commission (ARC) funded thirteen rural industry clusters under the rural JIAC program. Federal agencies provide financial and technical assistant to the 43 clusters and expect the clusters to transform their respective regions into high-growth economies stimulating formation of industries with high paid jobs. The interim findings of the evaluation of JIAC and AM-JIAC programs reported the clusters' progress towards the goals of the funding agencies (Mathematica Policy Research 2015).

2.4.3 Science/Technology/Business Parks

Science and technology parks in US are state/university driven clusters. By 2010, there were more than 170 such parks in the US which are the backbone of the US innovation system (Association of University Research Parks, AURP 2010). Stanford Research Park, originally established as Stanford Business Park in early 1950s, is the pioneer in the development of science parks. Frederick Terman, Dean of Engineering, first envisioned the potential of collaboration of Stanford University with industry and the City of Palo to generate income for university and the community. The university leased land to growing companies and the flood of corporations soon turned the region to Silicon Valley—the global capital of the technology world. Currently, the park homes over 150 companies including Hewlett-Packard (HP), Facebook and Tesla Motors (Stanford Research Park 2017). One of the reasons behind the success of Stanford Research Park is the support from the university initiatives that were carried out in parallel to the creation and expansion of the park. Honors Cooperative Program and Industrial Liaison Program are the two major initiatives that helped in the initial growth of the park. The Honors Cooperative Program enabled working professionals and engineers in electronics companies to enroll in graduate courses and stay updated with the current technology. This helped small companies to employ top talent by providing them training and continuous education in a fast-changing technological environment (Saxenian 1996). Also, through the Industrial Liaison Program, the selected company members (that guaranteed $5000 a year for 5 years) received access to Stanford's research projects, research results, and graduate students and had a chance to discuss their technical problems and possible solutions (US NRC 2013). Emerging industry focus, technology updated curriculum and courses, and start-up funds are the other initiatives that accelerated the success of the Stanford Research Park.

Named for the geographic area between three major research universities, Duke University in Durham, NC State University in Raleigh, and University of North Carolina in Chapel Hill, Research Triangle Park (RTP) is a prominent collaborative effort between the state and the region to initiate a high-tech R&D park. The idea of the research park originally came in 1952 as a vision to discontinue the migration of North Carolina's better college graduates traveling out-of-state for employment opportunities (Forbes 2012). The idea was to attract these graduates with high pay-ing jobs within the state. The state leaders realized the potential role of university cooperation for the park's success and developed strategies to use the research exper-tise of the universities mainly to emerging fields of pharmaceuticals, electronics and chemistry. Land for the park was initially acquired with a private land venture cre-ated by Pinelands Company. However, the development of the park was troubled due to the shortage of funding. The developers realized a possibility to raise more capital by redesigning the park in a public service orientation rather than for private gain. Funds were raised from wealthy and established North Carolinians motivated to contribute to the state's wellbeing. The funds were used to acquire the land that had been purchased by Pinelands and to transfer control of Pinelands to a non-profit Research Triangle Foundation formed in 1959 (Link and Scott 2003). Remarkable growth of the Park began in 1965 with the Federal announcement that the Park has been selected for its National Environmental Health Sciences Center. International Business Machines (IBM), which currently employs more than 10,000 professionals and staff members in RTP, also joined the park in the same year. Currently, the park homes 200 companies—employing 50,000 professionals and staff members and is the largest research park in the US (RTP 2017). The state's commitment towards the park over the past thirty years is the main factor behind its success that has trans-formed the state from one of the poorest in the southeastern US in 1960s to among the wealthiest in the region (US NRC 2013). Also, the park would not have been possible without the contributions of many ordinary North Carolinians who provided the initial capital.

Past experiences show that there is no single model guiding a path for the success of such parks, rather their performance is case specific. Felsenstein (1994) and Wallsten (2001) identify ambiguous performance of firms located in university-based science parks in the US. Most importantly, steady and long-term financial commitment from government and private sector and innovative culturing in the region help these parks in succeeding (Saxenian 1996). Most successful research parks tend to have a large research university or national laboratory at the core with a critical mass of highly trained knowledge workers supported by a strong public-private partnership among government, academia and industries.

2.4.4 Technology Transfer Offices

In US, the number of technology transfer intermediaries increased dramatically since 1980 after the enactment of the Bayh-Dole Act. Currently, most research universities in the US have their own TTOs. A survey of US universities' TTOs by Carlsson and Fridh (2002) showed that the majority of the TTOs follow similar technology transfer process which undergo a sequence of events of invention disclosure, review by TTO, application for IP right (patent, copyright, trademark), licensing, and/or start-up. Other than licensing and patenting, TTOs in recent decades are also increasingly focused on creation of spin-off firms (Siegel et al. 2007). According to Association of University Technology Managers (AUTM), the enactment of America Invents Act in 2011 shifted the US patent system emphasis to "first to file" system which has added responsibilities for TTOs operating most recently, to direct more careful evaluation for patentable inventions (AUTM 2015).

Universities have increasingly realized TTOs as source of generating revenue. According to the US Licensing Survey conducted by AUTM, TTOs established 450 new companies in 2002, totaling 4320 companies since 1980 (Fleischut and Haas 2005). More recently, AUTM US licensing Activity Survey reported the generation of 5145 US patents and formation of 705 startup companies with the US university TTOs, which are 9.5 and 5.1% more than that of 2011 respectively (AUTM 2015). The top three TTOs of the universities include Technology Licensing Office—Massachusetts Institute of Technology (MIT), Penn Centre for Innovation—University of Pennsylvania, and Centre for Technology Enterprise and Commercialization—Cornell University based on their generated revenue, disclosed inventions, issued patents, and successful spinoffs (GUV 2014).

2.4.5 Business Incubation Centers

Batavia Industrial Center, established in 1959, is the first incubator in the US (Wiggins and Gibson 2003; Hoffman and Radojevich-Kelly 2012; Mitra 2013). The center was started as an individual attempt of Mancuso family to gain from a big warehouse building that was purchased from a closed factory. However, business incubators did not get popularity until 1980s and only twelve business incubators were operating in US by that time (Carvalho 2015), most of which were established with a mission to revive the declining manufacturing industry. The actual growth of business incubation centers in US took place in the 1980s with strong support from government agencies, such as SBA. Furthermore, the emergence of technology-based firms in 1990s started a new trend for incubators, i.e., technology incubators. Each technology incubator focused on specific industrial and technological areas such as biotechnology, IT and environmental technology (Aernoudt 2004). The number of for-profit technology incubators increased dramatically in this period. These incubators were formed by private groups. Many of them failed within two years of their establishment due to lack of management and effective business counseling (Mitra 2013). Nevertheless, the incubators kept on evolving with new strategies and plans and many of them played important role in the development of competitive firms. According to National Business Incubation Association (NBIA), there were over 1250 incubators in the US at the end of 2012 (Carvalho 2015); about 90% of these incubators were the non-profit incubators (created and operated by the universities, government and non-profit organizations) while the rest were for-profit incubators (formed by private groups) (Badir Program for Technology Incubators 2012).

2.5   Support Programs

US support programs facilitating industry, academia and government collaborations are described in the following subsections.

2.5.1   Public Procurement

US government has supported evolution of number of high technology industries, such as semiconductors, computers, and aerospace, through procurement of products for defense and other public purposes. Use of Integrated Circuit (IC) chips by NASA in their different missions starting early 1960s, enabled the high technology companies in the US to improve revenues, and reduce costs through volume production (US NRC 2013). This also helped the high technology companies to find wider industrial and commercial applications of their technologies. Presently, the federal government is following similar programs in promotion of the green technology. US military plans to boost demand of the US advanced battery industry through purchases of electric vehicles for federal fleets. In 2010, the Advanced Vehicle and Power Initiative, led by US Army's Tank-Automotive Research, Development and Engineering Center (TARDEC), called for replacing 8% of the government truck fleet with electrified vehicles every year (TARDEC 2010). Also, the General Services Administration (GSA) plans to buy more than 40,000 alternative-fuel and fuel-efficient vehicles to replace aging and less-efficient fleets across federal agencies (US NRC 2012).

2.5.2 Tax Incentives

The US federal and state governments provide incentives for research and experimentation, domestic manufacturing, and early-stage financing to start-ups through various tax credit programs. The federal Research and Experimentation (R&E) tax credit provision established in 1981 is designed to stimulate national innovation by making research activities cheaper for businesses. The tax credit provision incentivizes research activities by reducing tax liabilities for organizations funding research, which lowers the after-tax cost of the research activities (Akabas and Collins 2014). This federal program was started as a temporary provision; the program expired and was renewed several times. Many companies and institutions raised concerns over the temporary nature of the program. Consequently, in 2015, the program was announced permanently. Provision of such tax incentive has been found beneficial to business R&D, particularly to stimulate increased R&D investment (Shapira and Youtie 2010).

The federal government has encouraged domestic manufacturing through tax deductions for consumer purchases of solar systems and electrified vehicles since mid-2000s (US NRC 2012). Around the same time, many state governments have introduced similar tax incentive programs to support emerging industries. Michigan state's Advanced Battery Tax Credits program provides partial refund of taxes to companies manufacturing battery cells and are engaged in advanced battery engineering (Shreffler 2010). Michigan state also has a Technology Collaboration Tax Credit program which provides incentives to small companies over business collaborations in emerging technologies (Michigan Economic Development Corporation 2009). New York and New Mexico have Solar Tax Credit program, which offer a generous tax incentive to those who purchase solar systems (New York Department of Taxation and Finance 2017). The federal and state tax incentives have strengthened the emerging technology industries in the US in one or many ways, by drawing private investment, empowering the small industries and leveraging domestic market.

2.5.3 Internship, Training, Entrepreneurship and Innovation in Curriculum

Universities and technical institutes in US have extensive internships and training programs with industries. These programs provide hands-on and on-the-job training to the students, which complement the academic training. Universities such as Belton University, Butler University and Cornell University are offering some of the best internship opportunities to their students. Most universities in the US offer academic programs in innovation and entrepreneurship to nurture and encourage entrepreneurial leaders. Entrepreneurship education was introduced in the US as early as 1940s (Zhou and Xu 2012). Also, from 2010, the federal government has emphasized the need of active participation between community colleges and industries in curriculum and workforce development (US MOE 2012).

2.5.4   Public-Private Partnership Programs

Importance of IUGP has been well realized in US from the early years of innovation, and the federal government/agencies have created several dedicated partnership programs. Some outstanding US partnership programs are Small Business Innovation Research program (SBIR), Small Business Technology Transfer program (STTR), Advanced Technology Program (ATP)/Technology Innovation Program (TIP), Grant Opportunities for Academic Liaison with Industry (GOALI), Partnerships for Innovation (PFI), and the cluster support programs. Federal agencies support these programs through grants and funding for research and experimentation, and through venture capital for start-ups, trainings and education.

2.5.4.1   SBIR

The SBIR program, first created in 1982 and coordinated by SBA, is dedicated to encouraging US small businesses to participate in federal R&D projects which can be commercialized. The program requires federal agencies with extramural R&D budget to award grants or research contracts to small businesses. Currently, there are 11 federal agencies participating in the program, which are required to allocate 3.2% of their annual R&D budget to SBIR (SBA 2017b). The SBIR program serves as a source of start-up and seed capital for innovations. Federal agencies also operate other supportive programs that help competitive recipients to get introduced to venture capital funds and other potential supports. In addition, state governments have emplaced programs to supplement the SBIR grants (US NRC 2013). Assessment of SBIR programs by National Research Council (NRC shows value creation through low-cost technologies and support for the fledgling entrepreneurs (US NRC 2008). According to US NRC (2012), SBIR is increasingly seen as a best practice around the world and the model has been adapted by several advanced and emerging economies with similar programs.

2.5.4.2   STTR

The STTR program, first created in 1992, considers the growing need to bridge the gap between basic science and commercialization of the results (SBA 2017b). The constitutional objective of STTR is to stimulate a partnership of ideas and technologies between innovative small businesses and research institutions through federally funded R&D (SBA 2014b). STTR requires federal agencies with extramural R&D budget to allocate 0.3% of the research budget to small businesses. Currently, there are five federal agencies participating in the STTR program. The assessment of STTR carried out by National Academy of Sciences, Engineering and Medicine (NASEM) showed that this program creates deeper and richer connection between small industries and research institutions, although many of the recipient industries found the program harder to manage in comparison to the SBIR program (NASEM 2016).

2.5.4.3 ATP/TIP

The ATP program, started in 1990, supports industries to accelerate the development activities of high-risk ventures, which eventually have higher potential payoffs (NIST 2005). By 2007, the program offered 824 technology commercialization awards to companies and joint ventures mainly in the areas of electronics and photonics, information technology, advanced materials, and biotechnology (Schacht 2011; Shapira and Youtie 2010). Evaluations of the program by NRC documented ATP to be an effective federal partnership program (US NRC 2001). For successful SBIR recipients, the program was increasingly valued as a source of early stage capital (US NRC 2012). Nevertheless, the program was ended in 2007 due to various difficulties in the reauthorization. NIST's TIP program was the successor of ATP and had similar mandate as latter; to accelerate innovation through high risk-reward research in areas of critical national need with targeted investments in transformational R&D (NIST TIP 2017). TIP funded 38 projects by 2010 (Schacht 2011); however, the program did not receive broad support from the US Congress and was also terminated in 2011.

2.5.4.4 NSF GOALI

Established in 1990, GOALI targets high-risk/high-gain research and funds small-group projects, fellowships and traineeships to stimulate interactions and exchange between universities and industries (NSF GOALI 2012; Martin-Vega et al. 2002). Such interactions increase the value of education and research, create an environment to discuss new exciting areas of research, and support industries providing a mechanism to leverage its research investment. Currently, there are 315 active awards under GOALI program. Larson and Brahmakulam (2002)reported GOALI as a key NSF funded program that has greatly expanded NSF's efforts to support collaborative research between universities, industry and government since 1990s.

2.5.4.5 NSF PFI

PFI, operational since 2000, is NSF funded program dedicated to regional innovation through partnerships among universities, industry, and local and regional government (Larson and Brahmakulam 2002). PFI has two complementary subprograms, Building Innovation Capacity, which funds advance technology research projects led by an interdisciplinary academic research team with at least one industry partner, and

Accelerating Innovation Research, which funds commercialization of previously NSF-funded research results with promising commercial potential (NSF PFI 2017). Currently, there are 42 active projects under PFI related to computers, science, and engineering discipline.

2.5.4.6   Cluster Support Programs

Federal agencies conduct different cluster support programs to manage the cluster initiatives through federal, state, and local governments, and universities and private sectors. US DOC's EDA started a support program as early as 2009 providing huge funding to the regional innovation clusters for their promotion and development (EDA 2016b). The Seed Fund Support (SFS) program, also known as Cluster Grants for Seed Funds of EDA, supports potential start-up companies (with commercialization potential) through the cluster-based seed capital fund. Cluster mapping project is a national program established to coordinate the regional innovation cluster initiative. The project is led by Harvard Business School's Institute for Strategy and Competitiveness in collaboration with EDA (2017). The US cluster mapping website was launched in 2014. The tool provides a platform to interact with a diverse group of stakeholders and facilitate decision making by providing a database of cluster initiatives and other economic indicators (Cluster mapping 2017).

A summary of the chapter is tabulated in Table 2.1.

Table 2.1 Summary of the IUGP trends and drivers in the US

	1960 and before	1960–1970	1970–1980	1980–1990	1990–2000	2000–2010	2010 and after
Governance	US DA, DARPA/US DOD, NIH, NASA, NIST, NSF, SBA	EDA/US DOC	US ED, US DOE, ETA/US DOL, SIA	SRC			
Legislations				Bayh-Dole Act, Stevenson-Wydler Act, Small Business Innovation Development Act, National Cooperative Research Act, Technology Transfer Act, Omnibus Trade and Competitiveness Act	American Technology Preeminence Act, National Technology Transfer Improvements Act, Small Business Technology Transfer Act	America COMPETE Act	Leahy-Smith America Invents Act
Intermediaries							
Research Institutes/Centers/Consortiums			IUCRC	ERC, STC,	MRSEC, FCRP Center	NSEC, NRI Research Center	The i6 Challenge: Proof of Concept Centers, NNMI Institute

(continued)

Table 2.1 (continued)

	1960 and before	1960–1970	1970–1980	1980–1990	1990–2000	2000–2010	2010 and after
Clusters						ATIP Cluster	ERIC, SBA's Pilot Contract based Cluster, JIAC, AM-JIAC, Rural JIAC
Science Parks/Business Incubators/ TTOs	University/State-Owned Science Park, Business Incubator			TTO			
Policies							
Public procurement		Public procurement of IC chips by NASA					Advanced vehicle and power initiative
Tax incentive				The federal Research and Experimentation (R&E) Tax Credit		Federal tax-credits for solar systems and plug-in electric vehicles, state-based advanced battery/technology collaboration/solar tax credit	

(continued)

Table 2.1 (continued)

	1960 and before	1960–1970	1970–1980	1980–1990	1990–2000	2000–2010	2010 and after
Academic Entrepreneurship and Innovation programs	Academic Entrepreneurship Program						
Partnership programs				SBIR	STTR, ATP	TIP, GOALI, PFI, SFS,	Emphasis on industry participation in curriculum development, US cluster mapping

References

Aernoudt, R. (2004). Incubators: Tool for entrepreneurship? *Small Business Economics, 23,* 127–135.

Akabas, S. & Collins, B. (2014). What is the research and experimentation tax credit? https:// bipartisanpolicy.org/blog/what-research-and-experimentation-tax-credit/. Accessed on May 20, 2017.

AURP. (2010). *The Power of Place 2.0-The Power of Innovation: 10 Steps for Creating Jobs, Improving Technology Commercialization, and Building Communities of Innovation.* Available at http://www.aurp.net/assets/documents/AURPPowerofPlace2.pdf.

AUTM. (2015). *The AUTM Briefing Book: 2015.* Available online in https://www.autmvisitors. net/sites/default/files/documents/AUTM%20Briefing%20Book%202015.pdf. Accessed on June 4, 2017.

Badir Program for Technology Incubators. (2012). *Business Accelerators and Business Incubators.*

Batterson, J. G. (2002). *Extending outreach success for the national nanoscale science and engineering centers—A handbook for universities.* Available online on http://www.nano.gov/node/ 625.

Ben-David, J. (1968). *Fundamental research and the universities.* Paris: OECD.

Blackwell, H. (2012). Emerging energy and intellectual property—The often unappreciated risks and hurdles of government regulations and standard setting organizations. *The National Law Review,* May 22, 2012.

Bradley, S. R., Hayter, C. S., & Link, A. N. (2013). Proof of concept centers in the United States: An exploratory look. *Department of economics working paper series,* The University of North Carolina Greensboro, May 2013.

Carlsson, B., & Fridh, A. C. (2002). Technology transfer in United States universities—A survey and statistical analysis. *Journal of Evolutionary Economics, 12*(1–2), 199–232.

Carvalho, L. (2015). *Handbook of research on entrepreneurial success and its impact on regional development.* IGI Global.

Chai, S., & Shih, W. (2016). Bridging science and technology through academic–industry partnerships. *Research Policy, 45,* 148–158.

Chubin, D. E., Derrick, E., Feller, I., & Phartiyal, P. (2010). *AAAS review of the NSF science and technology centers integrative partnerships (STC) program, 2000–2009.* Washington, DC: Final Report, AAAS.

Cluster mapping. (2017). http://www.clustermapping.us/about. Accessed on May 18, 2017.

Cornell University, INSEAD, WIPO. (2017). *The global innovation index 2017—innovation feeding the world.* Ithaca, Fontainebleau, and Geneva.

EDA. (2016a). https://www.eda.gov/archives/2016/challenges/jobsaccelerator/. Accessed on May 18 2017.

EDA. (2016b). https://www.eda.gov/archives/2016/news/blogs/2013/06/01/highlight.htm. Accessed on May 18, 2017.

EDA. (2017). https://www.eda.gov/about/cluster-mapping.htm. Accessed on May 18, 2017.

EDA The i6 Challenge. (2012). *The i6 challenge: proof of concept centers (fact sheet) 2012.* Available online in https://www.eda.gov/oie/ris/i6/2012/factsheet.htm.

Eurostat. (2017). R&D expenditure—Statistics explained. Available online http://ec.europa.eu/ eurostat/statistics-explained/index.php/R_%26_D_expenditure.

Felsenstein, D. (1994). University-related science parks—'Seedbeds' or 'Enclaves' of innovation? *Technovation, 14,* 93–100.

Fleischut, P. M., & Haas, S. (2005). University technology transfer offices: A status report. *Biotechnology Healthcare, 2*(2), 48–53.

Forbes. (2012). https://www.forbes.com/sites/davidkroll/2012/10/22/rtp-research-triangle-primer/ #64fc020a6a0b. Accessed on May 21, 2017.

Goldfarb, B., Henrekson, M., & Rosenberg, N. (2001). Demand versus supply driven innovations: US and Swedish experiences in academic entrepreneurship. *SIEPR Discussion Paper* No. 00–35, Stanford University.

GUV. (2014). GUV TTO and combined world rankings 2014. Available online in http://www.globaluniversityventuring.com/article.php/3919/guv-tto-and-combined-world-rankings-2014. Accessed on June 3 2017.

Hoffman, D. L., & Radojevich-Kelly, N. (2012). Analysis of accelerator companies: An exploratory case study of their programs, processes, and early results. *Small Business Institute Journal, 8*(2), 54–70.

Johnson, K. M. (2012). Build a competitive, low-carbon economy to secure America's energy future. In C. W. Wessner, Rapporteur (Eds.), *Clustering for 21st century prosperity: Summary of a symposium*. Washington, DC: The National Academic Press.

Larson, E. V., & Brahmakulam, I. T. (2002). *Building a New Foundation for innovation: Results of a workshop for the national science foundation*. RAND, Arlington, VA: Science and Technology Policy Institute.

Lewis, C. S. (2010). *Engineering research centers (innovations): ERC-generated commercialized products, processes, and startups*. SciTech Communications LLC.

Link, A. N., & Scott, J. T. (2003). The growth of research triangle park. *Small Business Economics, 20*(2), 167–175.

Manufacturing USA. (2017). https://www.manufacturingusa.com/institutes. Accessed on June 13 2017.

Martin-Vega, L., Seiford, L. M., & Senich, D. (2002). GOALI: A national science foundation university-industry liaison program. *Interfaces, 32*(2), 56–62.

Mathematica Policy Research. (2015). *Evaluation of the jobs and innovation accelerator challenge grants: Interim findings on multiagency collaboration and cluster progress (Interim Report)*. Princeton, NJ: Mathematica Policy Research.

Michigan Economic Development Corporation. (2009). Technology collaboration tax credit. http://files.meetup.com/1648509/Technology%20Collaboration%20Tax%20Credit.pdf. Accessed on May 21 2017.

Mitra, S. (2013). The problems with incubators and how to solve them. *Harvard Business Review*, 26 August 2013.

Mowery, D. C., & Sampat, B. N. (2005). The Bayh-Dole Act of 1980 and university–industry technology transfer: A model forother OECD governments? *Journal of Technology Transfer, 30*(1/2), 115–127.

NASEM. (2016). *STTR: An assessment of the small business technology transfer program*. Washington, DC: The National Academies Press.

NCSU. (2017). https://www.ncsu.edu/iucrc/. Accessed on May 13, 2017.

NEC, CEA, OSTP. (2011). *A strategy for american innovation: Securing our economic growth and prosperity*. Washington, DC.

New York Department of Taxation and Finance. (2017). https://www.tax.ny.gov/pit/credits/solar_energy_system_equipment_credit.htm. Accessed on May 21, 2017.

NIST. (2005). *Evaluation best practices and results: The advanced technology program*. Available online in http://nvlpubs.nist.gov/nistpubs/Legacy/IR/nistir7174.pdf.

NIST NRI. (2012). NRI to lead new five-year effort to develop post-CMOS electronics. Posted on 16 October 2012.

NIST TIP. (2017). https://www.nist.gov/technology-innovation-program. Accessed on May 21, 2017.

NNMI. (2016). *National network for manufacturing innovation program annual report*. Available online on https://www.manufacturingusa.com/resources/national-network-manufacturing-innovation-nnmi-program-annual-report.

NNN. (2017). http://www.internano.org/node/634. Accessed on June 11, 2017.

NSF NSEC. (2004). Nanoscale science and engineering (Program Solicitation). Posted online on 12 August 2004.

NSF GOALI. (2012). https://www.nsf.gov/pubs/2012/nsf12513/nsf12513.pdf. Accessed on May 13, 2017.

NSF STC. (2014). Science and technology centers: Integrative partnerships (Program solicitation). Posted online on 13 August 2014.

NSF ERC. (2015). Gen-3. Engineering research centers (ERC) partnerships in transformational research, education, and technology (Program solicitation). Posted on 24 July 2015.

NSF MRSEC. (2016). *Materials research science and engineering centers: Program solicitation.* Posted online on February 26, 2016.

NSF ERC. (2017). http://erc-assoc.org/. Accessed on July 12, 2017.

NSF IUCRC. (2017). https://www.nsf.gov/eng/iip/iucrc/home.jsp. Accessed on May 13, 2017.

NSF MRSEC. (2017). https://mrsec.org/mrsec-program-overview. Accessed on June 10 2017.

NSF PFI. (2017). https://www.nsf.gov/funding/programs.jsp?org=IIP. Accessed on May 13, 2017.

NSF STC. (2017). https://www.nsf.gov/od/oia/programs/stc/. Accessed on June 8, 2017.

Optimal. (2012). *The evaluation of the U.S. small business administration's regional cluster initiative (year one report).* MD: Optimal Solutions Group, LLC.

Rogers, J. D., Youtie, J., Porter, A., Shapira, P. (2011). *Assessment of fifteen nanotechnology science and engineering centers' (NSECs) outcomes and impacts: Their contribution to NNI objectives and goals.* NSF Award 0955089 Final Report, May 2011.

RTP. (2017). http://www.rtp.org/. Accessed on May 21, 2017.

Saxenian, A. (1996). *Regional advantage: Culture and competition in silicon valley and route 128.* Cambridge, MA and London: Harvard University Press.

SBA. (2014a). https://www.sba.gov/about-sba/sba-newsroom/press-releases-media-advisories/sba-announces-four-new-regional-innovation-cluster-awards. Accessed on May 18, 2017.

SBA. (2014b). *Small business technology transfer (STTR) program: Policy directive.* Available online in https://www.sbir.gov/sites/default/files/sbir_pd_with_1-8-14_amendments_2-24-14.pdf.

SBA. (2017a). https://www.sba.gov/content/innovative-economy-clusters. Accessed on May 18, 2017.

SBA. (2017b). https://www.sbir.gov/about. Accessed on May 18, 2017.

Schacht, W. H. (2011). *The TECHNOLOGY INNOVATION Program.* CRS Report for Congress, Congressional Research Service.

Shapira, P., & Youtie, J. (2010). The innovation system and innovation policy in the United States. Available in http://works.bepress.com/pshapira/19/. Accessed on May 15, 2017.

Shreffler, E. (2010). Michigan investments in batteries and electric vehicles. *National Academies Symposium on Building the U.S. Battery Industry for Electric Drive Vehicles,* Livonia, Michigan, 26 July, 2010.

SIA. (2010). Focus center research program: Government-industry co-sponsorship of university research. Available online on http://www.semiconductors.org/clientuploads/directory/DocumentSIA/Final%20SIA%20FCRP_OnePage_March2012.pdf.

SIA. (2012). Nanoelectronics research initiative: Government-industry partnership on university research. Available online on http://www.semiconductors.org/clientuploads/directory/DocumentSIA/Research%20and%20Technology/SIANRI0210.pdf.

Siegel, D. S., Veugelers, R., & Wright, M. (2007). Technology transfer offices and commercialization of university intellectual property: Performance and policy implications. *Oxford Review of Economic Policy, 23*(4), 640–660.

SRC FCRP. (2017). https://www.src.org/program/fcrp/about/mission/. Accessed on June 12, 2017.

SRI International. (2014). *The EDA i6 Challenge Program: Assessment & Metrics.* Final Report, 30 October 2014.

Stanford Research Park. (2017). http://stanfordresearchpark.com. Accessed on May 9, 2017.

TARDEC. (2010). *Advanced vehicle and power initiative.* Transportation Energy Security Team, TARDEC-National Automotive Center.

The World Bank. (2017). https://data.worldbank.org/. Accessed on December 17, 2017.

US Census Bureau. (2017). https://www.census.gov/quickfacts/fact/table/US/PST045216# viewtop. Accessed on December 17, 2017.

Us, N. R. C. (1987). *Science and technology centers: Principles and guidelines*. Washington, DC: National Academy of Sciences.

US NRC. (2001). *The advanced technology program: Assessing outcomes*. In C. Wessner (ed.), Washington, DC: The National Academies Press.

US NRC. (2003). *Securing the future: Regional and national programs to support the semiconductor industry*. In C. Wessner (ed.), Washington, DC: The National Academies Press.

Us, N. R. C. (2005). *The national science foundation's materials research science and engineering centers program: Looking back, moving forward*. Washington, DC: National Academies Press.

Us, N. R. C. (2008). An assessment of the SBIR program. In C. Wessner (Ed.), *Washington*. Washington, DC: The National Academies Press.

US DOE. (2010). Fiscal year 2010 energy efficient building systems regional innovation cluster initiative. *Funding Opportunity Announcement*.

US DA. (2011). USDA, Agricultural Research Service (ARS): Technology transfer program and formal links to economic development through its Agricultural Technology Innovation Partnership Program (ATIP). *USDA Technology Transfer by ARS*.

Us, N. R. C. (2011b). *Growing innovation clusters for american prosperity: Summary of a symposium*. Washington, DC: The National Academic Press.

Us, M. O. E. (2012b). *Integrating industry-driven competencies in education and training through employer engagement*. Washington, DC: Office of Vocational and Adult Education.

US NRC. (2012). *Rising to the challenge: US Innovation Policy for the Global Economy*. Washington, DC: National Academies Press,

US NRC. (2013). *Best practices in state and regional innovation initiatives*. In C.W. Wessner (ed.), Washington, DC: The National Academic Press.

US ED, OII. (2017). https://innovation.ed.gov/. Accessed on May 7, 2017.

Wallsten, S. (2001). The role of government in regional technology development: The effects of public venture capital and science parks. *Stanford University: SIEPR Working Paper*.

Wiggins, J., & Gibson, D. V. (2003). Overview of US incubators and the case of the Austin Technology Incubator. *International Journal of Entrepreneurship and Innovation Management, 3*(1/2), 56–66.

Zhou, M., & Xu, H. (2012). A review of entrepreneurship education for college students in China. *Administrative Sciences, 2*, 82–98.

Chapter 3
Case Study: Norway

Abstract The innovation system and IUGP settings of Norway are of special interest because the country has transformed its oil and gas-based growth to knowledge-based development. Norwegian example is most suitable for the small-sized wealthy nations with abundant hydrocarbon resources which are committed to transform their natural capital to knowledge capital. This chapter provides a detailed account on the past and present of innovation system and IUGP initiatives that helped Norway in developing its knowledge ecosystem. We explore the cultural and institutional settings in Norway to find the needs and motivations to transform the Norwegian hydrocarbon economy to knowledge-based economy. The role of the state government has been remarkable in establishing the sector-based public research institutes and a collaborative framework between the research institutes, higher education institutes, and industry. In addition, the enactment of concession laws and the law on the right to inventions made by employees motivated the knowledge-based workforce to innovate and commercialize. At the same time, the establishment of science parks and business parks helped the young entrepreneurs and SMEs to increase their competence level and to compete at the national and international levels. Also, the decentralized public procurement helped in stimulating the domestic market for innovation. Lastly, the public-private partnership programs created further opportunities for technology transfer and commercialization.

3.1 Background

Norway has an economy that is largely based on its natural resources. Until 1960s, Norway was identified as one of the economically weakest countries in Europe. By 1970, Norway had caught up with most Western European countries and by 2001, Norway became one of the richest countries in the world (Engen 2009). As of 2016, Norway had a population of 5.23 million and a GDP per capita (in Purchasing Power Parity) of 59,384 International Dollars—11th highest in the world (The World Bank 2017). Norwegian government plays a central role in the technological development

---

R. Pradhananga co-authored this chapter.

© Springer Nature Switzerland AG 2020

W. Nawaz and M. Koç, *Industry, University and Government Partnerships for the Sustainable Development of Knowledge-Based Society,* Management and Industrial Engineering, https://doi.org/10.1007/978-3-030-26799-5_3

of the country. Identification of oil and gas deposits in 1960s brought a major change to the Norwegian economy. From early 1980s, the government invested revenue collected from oil and gas sector to research and development and emphasized collaborations between academia, public research institutes, and industries. The country is making increasing investments in R&D; its R&D expenditure in 2016 was 2.04% of the GDP, and business and higher education sectors accounted for 53% and 33% of this expenditure, respectively (Nordic Institute for Studies in Innovation, Research and Educator NIFU 2017). Consequently, the 2017 GII of Norway was 53.14—19th highest among the 127 countries across the world (Cornell University, INSEAD, and WIPO 2017).

3.2 Institutional and Cultural Setting

Innovation in Norway is different from innovation in other parts of the world since Norwegian innovation stem from the abundant natural resources rather than the high-tech industries as seen in the case of US. The government of Norway has a key role in the entire innovation system. The government has taken several strategic decisions to use the resources in a way that benefits the domestic industries. Historically, Norway relied on fishing, agriculture, shipping and other related industries for its economic growth—based on the geography of the country. This provided opportunities for development of the small-scale industries in Norway, most remarkably in the shipping sector. Earlier in twentieth century, Norway recognized the possible opportunities with its mountainous terrain for mining and production of hydroelectric power. Huge investments were made on energy-related sectors, such as the electrometallurgical and chemical industries, which formed the basis of a group of centralized large industries in Norway. By this time a few public research institutes were also formed to help the technological needs of the industries. Government efforts to nationalize the ownership of the natural resources through concession law supported decentralization of the industry and economic growth with continuous local support.

The development of the oil and gas sector opened up a huge market for domestic manufacturing and service firms. The government-introduced concessions, tax system and other regulations to foster the Norwegian technological capacity—mainly in the field of engineering, Information and Communication Technology (ICT), business services related to oil and gas sector, and the shipbuilding sector. Revenue from oil and gas sector was invested in R&D and higher education (Smith et al. 1996). Traditional industrial sectors, such as fishing, shipping, electrometallurgical and chemical industries, were also technologically improved since these remained important source of income and employment in some regions of the country (Fagerberg et al. 2009).

Public research institutes (PRIs) played a significant role in the internationalization of research and innovation in Norway (OECD 2017a). In early years of innovation, smaller high-tech industries relied on PRIs for their research and innovation. Norwegian Higher Education Institutes (HEIs), which comprise of universities and state-owned university colleges, paced slightly slower because the research institutes and HEIs were heterogeneous and fragmented. The government incentivized mergers between the HEIs and among HEIs and PRIs to improve the quality of research and competence of scholars (OECD 2017a). During the same period, the government introduced policies to promote public-private cooperation between HEIs, PRIs, and industries to raise the research and innovation competence of the country to international standards. As a result, 1980s mark a strong increase in formal research collaborations in Norway. Afterwards, most large industries established strong research competency in their sectors, especially in the oil and gas sector.

One unusual feature of the Norwegian innovation system is that Norway has shown a good growth in productivity and income with comparatively low R&D investment (Fagerberg et al. 2009). A possible reason behind could be the Norwegian system to work collaboratively which helps to spill the research and innovation cost over different sectors. In recent years, public-private funding to support business innovation has increased substantially. The main funders are Ministry of Education and Research, Ministry of Trade and Industry (NHD), and Ministry of Health and Care Services (HOD). Research Council of Norway (RCN), formed in 1993, the Industrial Development Corporation of Norway (SIVA), formed in 1968, and Innovation Norway (IN), established in 2003, offer strategic programs and incentives to support innovation and commercialization. RCN operates under Ministry of Education and Research and is formed as a merger of five former research councils for Scientific and Industrial Research (NTNF), Agriculture (NLVF), Science and Humanities (NAVF), and Fisheries (NFFR) and Applied Social Sciences (NORAS).

Institutional buildings in Norway have a strong foreign influence (Gulbrandsen and Nerdrum 2009a). Over the last two decades, development of intermediary structures, science parks, business incubators, TTOs and regional clusters has received substantial focus (OECD 2017a). The recent programs have increased emphasize on the basic research and the high-risk high-gain research of international standards (Larédo and Mustar 2001).

3.3 Legislations/Regulations

Legislations that shaped the current Norwegian innovation system started as early as 1906, a year after the dissolution of the union between Norway and Sweden. Concession Laws passed at that time by the Norwegian government regulated the foreign ownership of hydropower; since the Norwegian economy required foreign investments time to time for exploitation of their natural resources, mainly water at that time, the government strategically granted rights to foreign companies to make such investments but each time with negotiations to work with local industries and organizations (Hanisch and Nerheim 1992; Engen 2009; Ville and Wicken 2013). Also, the development work carried out by the private firms was to be handed back to the government after a period of 60–80 years, with no additional compensation to the owners. Ultimately, these laws proved to be the major success for the Norwegian domestic industries.

Technology Agreement, or Goodwill Agreement, introduced in 1979 helped Norway to build significant domestic research capacity in oil and gas sector (Engen 2009). The agreement provided incentives to foreign companies for partnering and contracting with the Norwegian firms and research institutes. It also rewarded R&D spending and knowledge transfer to domestic organizations. Introduction of Skattefunn Scheme warranted by Taxation Law in 2002 declared tax credits to R&D-related costs (European Commission 2014), and this facilitated R&D spending largely in all sectors. Furthermore, amendments to the 'Law on the right to inventions made by employees' in 2003 and 'University and university colleges act' of 2005 significantly strengthened the commercialization activities of the Norwegian HEIs. The former has given universities and university colleges the rights to the IP created as a result of research (Gulbrandsen and Nerdrum 2009b). This provided a simple yet comprehensive structure for commercialization and value creation from public R&D in Norway.

3.4   Intermediary Structures

Following are the key intermediary structures that facilitated IUGPs in Norway:

3.4.1   Government/Semi-government Research Institutions

Norwegian innovation system witnessed two groups of public research organizations which played a fundamental part in the knowledge-based development of the country: (i) the PRIs; and (ii) the innovation centers, including Centers of Excellence (CoE), Centers for Research-based Innovation (CRI), Centers for Environment-friendly Energy Research (CEER) and Norsk Katapult Centers. The latter have been formed under the RCN and/or IN and/or SIVA's support programs.

3.4.1.1 PRI

Norwegian innovation is characterized by strong sector-oriented PRIs which have a history starting from the 19th century when only a few technical-industrial institutes were established in agriculture and marine sectors. The main growth of PRIs occurred after the World War II (Gulbrandsen and Nerdrum 2009a). NTNF with support from NHD created several multi-disciplinary, industry-oriented PRIs in 1950s. The Central Institute of Industrial Research (Sentralinstituttet—SI) is one of the most influential PRIs owned by NTNF. Until mid-1980s, ministries and research councils owned the PRIs. Since then, most institutes became autonomous and operated as foundations and non-profit organizations (OECD 2017a). Norwegian PRIs are fragmented and diverse in terms of the size, scientific orientations and financial sources (OECD 2017a). While some institutes are large, multi-disciplinary and with several employees, many are small, specialized with few employees. The Foundation for Scientific and Industrial Research (Stiftelsen for industriell og teknisk forskning—SINTEF) established in 1950 by Norwegian Institute of Technology (now part of Norwegian University of Science and Technology, NTNU) is currently the largest PRI in Norway.

PRIs played a prominent role in research and innovation during 1970s and 1980s due to the increasing technical and innovation demands of industries, mainly from large industries in oil and gas sector. Research activities of these institutes were focused on the applied sciences, and particularly to the fields of engineering, technology and natural sciences. Over time, PRIs went through changes in their organizational structure and scientific orientation. At present, PRIs are supporting the business innovation in two distinct ways. PRIs provide basic infrastructure of R&D which fulfills industrial needs for knowledge, competence and equipment. Additionally, it act as a key intermediator between firms and universities (Gulbrandsen and Nerdrum 2009b).

PRIs operate with private, national, and international funds. A huge share of funding comes as block funding from RCN while all PRIs can also apply for competitive national grants under support programs from RCN. PRIs have shown a strong participation in these programs as well as in other EU funding programs (OECD 2017a). As of 2017, there are 100 PRIs in Norway, of which 44 PRIs can receive block funding from the government. The rest of the PRIs are publicly owned research centers directly funded by the ministries and public institutions with specific public missions. While most industries have valued PRIs as important partners for their research needs (RCN 2015a), the PRIs have also shown significant impact on knowledge transfer through licensing, patenting, and spin-off activities (RCN 2015b).

3.4.1.2   CoE

CoEs are multi-institutional centers, first established in 2001 by the RCN. The CoE program encourages collaborations on basic research-driven areas (OECD 2017a). The centers are led and hosted by universities and in some cases by the PRIs. RCN provides funds to the centers for a maximum period of ten years. The centers also receive substantial financial contributions from the partners, mainly from the host institution. By 2017, the program funded 34 centers, of which 13 centers from round 1 (2002) ended in 2012. Currently there are 21 centers operational from round 2 (2007) and round 3 (2012) (OECD 2017a). Many researchers from CoE programs were successful in obtaining the European Research Council grants (RCN, 2016a). Evaluation of the CoE program carried out in 2010 confirmed effectiveness of CoE centers in strengthening internationalization of the Norwegian research (Langfeldt et al. 2010).

3.4.1.3   CRI

RCN's Centers for Research-based Innovation encourages active cooperation between industries and research groups for scientific research with high potential for innovation and value creation. Introduced in 2007, the program provides a long-term funding of up to eight years for the establishment of the centers in industry-oriented research (RCN 2008). The centers can be hosted by a university, university college, PRI, R&D performing company or a public service provider. So far, the program has established a total of 38 centers; 14 centers in 2007, 7 in 2011 and 17 in 2015. SINTEF and NTNU host more than half of these centers. Evaluation of the centers carried out in 2010 indicate a success in stimulating research-based innovation through CRIs (RCN 2010).

3.4.1.4   CEER

CEERs, first in place since 2009, are established under RCN. The aim of the centers is to fund scientific research in selected areas of environment-friendly energy and carbon capture and storage (CCS) technology, which have a high potential of innovation, commercialization and value creation. The centers facilitate active collaboration between dynamic research groups, public sector, and user partners from trade and industry (RCN 2017a). The host institution of the center can be a university, university college, or a research institute. RCN funds the centers for a maximum period of eight years. The centers also receive substantial financial support from the host institution and the collaborative partners. So far, RCN has established 11 centers under this program, and 8 more centers are planned to be launched in the future (RCN

2017a). The centers cover a wide range of areas in environment-friendly energy such as the hydropower, smart grids, energy efficiency in trade and industry, CO_2 management, solar cells, biofuels and zero-emission urban zones (RCN 2016b). Mid-term evaluation of the eight CEERs established in the first round has reported that the centers are showing outstanding performance in terms of scientific value and training (RCN 2013).

3.4.1.5 Norsk Katapult Center

Norsk Katapult centers aim to bring the Norwegian industry close to the research institutions through multipurpose centers (SIVA 2017a). The centers are co-created by SIVA, RCN and IN and is administered by SIVA. The centers provide easy access to testing, simulation, and visualization facilities to foster innovation with faster and cheaper translation of new concept to a marketable outcome.

3.4.2 Government Initiated Innovation Clusters

Norwegian government initiated a comprehensive innovation cluster support program in 2014 which was built upon the earlier cluster programs. The program aimed to enhance the active collaborations between enterprises, mainly SMEs, research communities, and the public sector, to increase the innovation potential and competence of the individual industries through cluster dynamics (IN RCN SIVA 2015). Financed by the Ministry of Trade and Industry and Ministry of Local Government and Modernization, this program is collaborative effort between IN, SIVA and RCN. The support program includes: Arena clusters, Norwegian Centers of Expertise (NCE), and Global Centers of Expertise (GCE).

3.4.2.1 Arena Clusters

Arena is a national cluster program established in 2002 to provide financial and professional support to regional business communities that have high concentration of companies in an area of expertise (IN RCN SIVA 2015). The program aims to promote business innovation through collaboration between businesses, R&D institutions, and public sector. The support offered by the government is for three years which is extendable to another two years based on the circumstances. By 2015, the program supported more than 70 clusters (IN RCN SIVA 2015) and currently there are 19 clusters operational under this program (Arena 2017). The clusters in general are in an early phase of organization and are referred as immature clusters. These clusters can be small or large and can have regional, national and international participants. Nevertheless, the evaluation of the Arena Clusters program highlights less emphasis on the link between research and industry—different from the cluster programs in other nations (Jakobsen and Røtnes 2012). For many Arena clusters, the role of universities, university colleges and the knowledge sector are peripheral. However, for sectors such as maritime, which are primarily customer driven, the effect of such weak linkage to knowledge appeared insignificant.

3.4.2.2   NCE

Started in 2006, NCE is a long-term cluster program that aims to establish mature business clusters with a national position (IN RCN SIVA 2015). The program supports dynamic business clusters that have developed through a high level of interactions and collaborations between business entities and have potential to grow in national and international market. Financial and technical supports are provided for up to ten years. As of 2017, there are 14 NCE clusters receiving such supports (NCE 2017). Evaluation of NCEs highlighted relatively small knowledge component in some NCE cluster projects (Jakobsen and Røtnes 2012). However, the program has positively impacted the knowledge ecosystem in Norway through increased collaborations between the public, industry and knowledge partners.

3.4.2.3   GCE

The GCE program, established in 2014, aims to establish mature clusters with global positioning (IN RCN SIVA 2015). The program funds business clusters with dynamic cooperation in strategic areas and with international competence level in R&D. The GCE clusters comprise of technology and market leaders that are part of global knowledge and value chain (GCE 2017). To strengthen the ongoing innovation and internationalization, the program focuses on cluster management, knowledge links, innovative collaborations, and cluster to cluster collaboration (IN, RCN, SIVA 2015). Currently there are 3 GCEs operational under this program. The clusters receive funding for a maximum of ten years and are additionally supported through the R&D budget of the participants.

3.4.3 Science/Technology/Business Parks

State enterprise SIVA has a central role in the establishment of science parks in Norway. SIVA is responsible for the provision of government investment in establishing the science parks. As a result, the partial ownership of the developed parks remains with SIVA. SIVA also administer an industrial program to establish business gardens/parks. The only difference between science/research park and business garden/park is that the latter represents cooperation between the knowledge-based companies in a region (SIVA 2017b).

Oslo Science Park established in 1986 and Tromsø Science Park established in 1990 are the inaugural science parks in Norway. Following suite with the Bayh Dole Act, Norway passed the 'Law on the right to inventions made by employees' in 2003. The enactment of this law increased the cooperation between universities and university colleges, which were using the science parks before, with the industries (European Commission 2012). According to Rotefoss et al. (2010), business gardens have shown positive impacts in the business development in Norway. By 2017, SIVA's innovation program supported 39 business gardens and 25 science and research parks (SIVA 2017b).

3.4.4 Technology Transfer Offices

Establishment of TTOs in Norway started effectively after the enactment of the law on the right to inventions made by employees. The interest in TTOs further developed when universities were authorized and funded to have their own TTOs. Currently, there are 8 TTOs affiliated with academic institutions in Norway (RCN 2011). A survey of Norwegian TTOs carried out in 2003 indicates a strong orientation of the TTOs towards commercialization compared to patenting and licensing (Gulbrandsen and Nerdrum 2009b). RCN's support program, Commercializing R&D Results (FORNY), is one of the main funding sources for commercialization activities in TTOs.

3.4.5 Business Incubation Centers

Similar to science parks and TTOs, business incubators in Norway proliferated in early 2000s. Currently, Norway comprises of a national network of business incubators established with support of SIVA. Similar to the science parks, SIVA has a partial ownership in 43 business/industry incubators. All incubators have a regional focus and are linked to regional universities, science parks, innovation centers, and industries (Cleantech Scandinavia 2009). SIVA's incubation program assures quality of the services offered to the entrepreneurs by the incubators (SIVA 2017b). As of 2017, 45 incubators received basic funding support through SIVA.

3.5 Support Programs

Norwegian public-private sectors have initiated several support programs to promote collaborations between industry, academia and government as discussed in the following subsections.

3.5.1 Public Procurement

In recent years, Norwegian government has increasingly realized the role of public procurement in stimulating the domestic market for innovation. The government has deployed a 'procurement for innovation plan' since 2013 to decentralize the public procurement (OECD 2017b). Public Management of e-Governance (DIFI) and the National Program for Supplier Development helped in developing a national method for procurement of innovation. NHD and a number of ministries, agencies and municipalities are developing policies to encourage and implement innovative public procurement programs at a national level. Green public procurement that supports green development in the country is one of the earliest efforts in this regard. From 2001, the Norwegian Public Procurement Act requires state counties and local authorities to consider environmental impacts of procurement activities. In 2009, approximately 70% of the public procurement passed the environmental requirements in the procurement process. Since 2010, DIFI is providing information to municipalities and counties to integrate environmental requirements in public procurement (Simanovska 2013).

3.5.2 Tax Incentives

As a commitment to make investments in R&D, Norwegian government introduced R&D tax credit program, SkatteFUNN in 2002. Under the program companies involved in R&D projects can qualify for tax credit that comes in the form of possible deduction in the company's payable corporate tax. In 2002, only SMEs could benefit from the tax credit, however, from 2003, large firms were also able to participate in the program (European Commission 2014). To be eligible to participate in the program: (i) the R&D project must generate new knowledge; and (ii) skills and capabilities must be oriented for the development of new/improved products, services or methods (SkatteFUNN 2014). Ministry of Trade administers the program and RCN, IN and the Directorate of Taxes approve the tax claims together. As of 2017, SMEs can receive 20% tax reduction of the R&D cost, whereas the same can be 18% for the large enterprises. By 2013, the program approved 24,619 tax claim proposals from 10,250 companies (European Commission 2014). A study on the impact of the program during 2002–2006 showed an increase in the R&D expenditure the enterprises in Norway (Hægeland and Møen 2007). The impact was significant for small, low-tech and low-skilled firms.

3.5.3 Internship, Training, Entrepreneurship and Innovation in Curriculum

Universities, colleges, and technical institutes in Norway offer wide range of internships and training programs with industries. NTNU and University of Oslo are some of the institutions with best internship programs. Most universities also offer specialized academic programs in innovation and entrepreneurship. Most engineering schools in Norway had entrepreneurship courses by mid 1980s (European Commission 2008). Most vocational training institutes structure training programs under industrial advisory, and more recently, universities are also engaging with industry leaders to align the academic curriculum with industrial needs.

3.5.4 Public-Private Partnership Programs

Norway had collaborations between industry and research institutions from its earlier years of innovation. However, until late twentieth century these collaborations were limited between industries and PRIs mainly for applied and user-controlled research. The struggle of enterprises to gain access to national and international competence combined with the enhanced capabilities of the universities and the high-quality research outcomes opened up the opportunities for technology transfer, commercialization, and public private partnership. Following subsections present the initiatives taken by the Norwegian government to support innovation-oriented research through public private partnership, including FORNY, User-driven Resource based Innovation (BIA), Regional R&D and Innovation (VRI), large scale programs, start-up supports and Industrial Research and Development (IRD).

3.5.4.1 FORNY

FORNY is a joint research program between RCN and IN dedicated for the commercialization of research-based business ideas with considerable market potential

(RCN 2009). It started in 1995 as a joint project and was later established as a program. FORNY aims to strengthen research-based innovation in Norway through active cooperation between research institutions, entrepreneurs, investors, industry, and government authorities. The program provides support for establishment of professional commercialization units such as the TTOs, science parks, incubators. The program offers financial support in four categories: infrastructure, commercialization, verification of technology, and scholarships. By 2009, the program supported approximately 300 start-ups and employed around 700 professionals. Evaluation of FORNY projects carried out by Borlaug et al. (2009) highlighted a lack of focus of the program on TTOs since the commercialization units did not work well. The authors called for a high level of engagement between the members of program to ensure broader transfer of knowledge.

3.5.4.2   BIA

BIA is a user-driven research-based innovation program administered by RCN. The program was launched in 2005 and is targeted at Norwegian trades and industry (RCN 2014) seeking to become more innovative and competitive. It promotes active collaboration between industries and research groups, and among companies at national and international level. The program partially funds the R&D project of the industry without any thematic and sectorial constraint. It expects the funded projects to generate significant new R&D initiatives in Norwegian trade and industry which create value in the long-term (RCN 2014). Currently BIA is recognized as a key funding program for industry-oriented research. The evaluation of the BIA project confirmed that the program has both short-term and long-term potential benefits (Mahieu et al. 2012). In short-term it can increase employment while in long-term it has a potential to benefit Norwegian industry and society as a whole.

3.5.4.3   VRI

VRI is RCN's program for regional R&D and innovation in Norway. The program was launched in 2007 for a time frame of ten years (RCN 2012a). VRI has objective to promote innovation, knowledge development, and value creation through regional collaboration (RCN 2004). Regional partnership, which consists of representatives from trade and industry, R&D institutions, public authorities and other funding agencies can apply for the professional and financial support from VRI program. The partnership must contribute to the 50% of the project funding. The program activities predominantly occur in the 15 VRI regions in Norway. The program supports innovation focused and industry-oriented projects.

3.5.4.4 Large-Scale Programs

RCN's large-scale programs promote concentrated and integrated research activities to achieve long-term knowledge and competence building in technological areas of societal value (RCN 2012b, 2017b). The programs support ten-year thematic projects which are funded by the industries, academia, public sector, and civil society. The programs have priorities for structuring the research efforts targeted for industry involvement, user participation, and interdisciplinary and international cooperation. The prioritized programs under seven themes include: agriculture, marine, industry and health; sustainability and marine; nanotech and advanced materials; climate change; petroleum; energy; and ICT (OECD 2017a).

3.5.4.5 Start-up Supports

IN provides financial support, advisory services, and promotional services to entrepreneurs, young companies, and SMEs with growth potential and an established innovative business concept (IN 2017). Ultimately, such a program supports and promotes innovation at regional and national level. Some particular assistance provided by IN under this program includes IPR and commercialization advisory, linkage to investors network, monitoring service and start-up grants.

3.5.4.6 IRD

Established in 1994, IRD is a support program administered by IN with an objective to promote development of new products, services and solutions for creating high value in national and international market (European Commission 2010; IN IRD 2013). The program provides grants to Norwegian SMEs for developing new products or services in demand of a pilot customer which could be a private company (foreign or Norwegian) or a public entity such as a government agency, a hospital or a municipality. A typical IRD project runs for one to three years. Evaluation of IRD projects in 2014 highlighted technological success of more than 80% of the projects (European Monitoring Center on Change EMCC 2016).

A summary of the chapter has been tabulated in Table 3.1.

Table 3.1 Summary of the IUGP trends and drivers in Norway

	1960 and before	1960–1970	1970–1980	1980–1990	1990–2000	2000–2010	2010 and after
Governance	Ministry of Education and Research, NTNF/NHD, NLVF, NAVF	SIVA	NFFR	NORAS	RCN/Ministry of Education and Research	IN, HOD	
Legislations	Concession Law		Technology or Goodwill Agreement			SkatteFUNN Scheme Warranted by Taxation Law, Law on the Right to Inventions Made by Employees, University and University Colleges Act	
Intermediaries							
Research institutes/centers/consortiums	PRI					CoE, CRI, CEER	Norsk Katapult Center
Clusters						Arena Clusters, NCE	GCE
Science parks/business incubators/TTOs			Science Park		Business Gardens	TTOs, Incubators	

(continued)

Table 3.1 (continued)

Policies	1960 and before	1960–1970	1970–1980	1980–1990	1990–2000	2000–2010	2010 and after
Public procurement						Green Public Procurement	Procurement for Innovation
Tax incentive						SketteFUNN R&D Tax Credit Program	
Academic entrepreneurship and innovation programs				Academic Entrepreneurship Program			
Partnership programs					FORNY, IRD	BIA, VRI, Large Scale Programs (FUGE, HAVBRUK, NANOMAT, NORKLIMA, PETROMAK, RENERGI, VERDIKT), Start-UP Supports to SMEs	Large Scale Programs (BIOTEK2021, HAVBRUK2, NANO2021, KLIMAFORSK, PETROMAKS2, ENERGIX, IKTPLUSS)

References

Arena. (2017). http://www.arenaclusters.no/the-arena-programme/arena-klyngene/. Accessed August 15, 2017.

Borlaug, B., et al. (2009). *Between entrepreneurship and technology transfer: Evaluation of the FORNY programme* (Rapport 19/2009. Nifu-Step).

Cleantech Scandinavia. (2009). *Nordic science parks + incubators* (Incubator Report-2). Cleantech Scandinavia Malmö, February 2009.

Cornell University, INSEAD, & WIPO. (2017). *The global innovation index 2017—Innovation feeding the world.* Ithaca, Fontainebleau, and Geneva.

EMCC. (2016). *Norwegian industrial research and development contracts (IRD).* Available at https://www.eurofound.europa.eu/observatories/emcc/erm/support-instrument/norwegian-industrial-researchand-development-contracts-ird.

Engen, O. A. (2009). The development of the Norwegian petroleum innovation system: A historical overview. In J. Fagerberg, D. C. Mowery, & B. Verspagen (Eds.), *Innovation, path dependency and policy. The Norwegian case* (pp. 179–207). Oxford: Oxford University Press.

European Commission. (2008). *Entrepreneurship in higher education, especially within non-business studies* (Final Report of the Expert Group). European Commission.

European Commission. (2010). *European enterprise awards 2010 shortlist.* Available at http://europa.eu/rapid/press-release_MEMO-10-77_en.htm?locale=en.

European Commission. (2012). *Erawatch country reports 2012: Norway.* Available at https://ec.europa.eu/jrc/en/publication/erawatch-country-reports-2012-norway.

European Commission. (2014). *A study on R&D tax incentives. Annex: Good practice cases* (Final Report). Available at https://ec.europa.eu/taxation_customs/sites/taxation/files/resources/documents/taxation/gen_info/economic_analysis/tax_papers/good_practice_cases.pdf.

Fagerberg, J., Mowery, D. C., & Verspagen, B. (2009). The evolution of Norway's national innovation system. *Science and Public Policy, 36*(6), 431–444.

GCE. (2017). http://www.gceclusters.no/the-arena-programme/. Accessed August 15, 2017.

Gulbrandsen, M., & Nerdrum, L. (2009a). Public sector research and industrial innovation in Norway: A historical perspective. In J. Fagerberg, D. C. Mowery, & B. Verspagen (Eds.), *Innovation, path dependency and policy: The Norwegian innovation system.* Oxford, UK: Oxford University Press.

Gulbrandsen, M., & Nerdrum, L. (2009b). University-industry relations in Norway. In J. Fagerberg, D. C. Mowery, & B. Verspagen (Eds.), *Innovation, path dependency and policy: The Norwegian innovation system.* Oxford, UK: Oxford University Press.

Hægeland, T., & Møen, J. (2007). *Input additionality in the Norwegian R&D tax credit scheme* (Reports 2007). Statistics Norway.

Hanisch, T. J., & Nerheim, G. (1992). Norsk oljehistorie. Frau vantro til overmot? Norsk Petroleums-forening, Leseforeningen, 1.

IN. (2017). http://www.innovasjonnorge.no/en/start-page/our-services/start-ups/. Accessed August 28, 2017.

IN IRD. (2013). http://www.innovasjonnorge.no/contentassets/68685a9f5fb6498581893886792951f8/2013-02-ird-fact-sheet.pdf. Accessed August 28, 2017.

IN RCN SIVA. (2015). *Programme description Norwegian innovation clusters.* Available at http://www.innovasjonnorge.no/contentassets/b80bc957214b487cb54fa1b927dbf40d/programme-description-en-12.01.15.pdf.

Jakobsen, E., Røtnes, R. (2012). *Cluster programs in Norway: Evaluation of the NCE and arena programs.* Menon-Publication.

Langfeldt, L., Borlaug, S. B., & Gulbrandsen, M. (2010). *The Norwegian Centers of Excellence Scheme. Evaluation of Added Value and Financial Aspects* (Nordic Institute for Studies in Innovation, NIFU SETP Report). Oslo.

Larédo, P., & Mustar, P. (2001). *Research and innovation policies in the new global economy.* Cheltenham: Edward Elgar.

Mahieu, B. Arnold, E., Horvath, A., & Rosemberg, C. (2012). *Evaluation of the research council of Norway*. Technopolis-Group.

NCE. (2017). http://www.nceclusters.no/about-nce/. Accessed August 15, 2017.

NIFU. (2017). https://brage.bibsys.no/xmlui/bitstream/handle/11250/2473523/Nordic%20newsletter%202016.pdf?sequence=1. Accessed December 20, 2017.

OECD. (2017a). *OECD reviews of innovation policy: Norway 2017*. Paris: OECD Publishing. https://doi.org/10.1787/9789264277960-en.

OECD. (2017b). *Public procurement for innovation: Good practices and strategies*. Paris: OECD Publishing. https://doi.org/10.1787/22190414.

RCN. (2004). https://www.forskningsradet.no/prognett-vri/Programme_Description/1224529235302. Accessed August 28, 2017.

RCN. (2008). https://www.forskningsradet.no/prognett-sfi/About_the_SFI_scheme/1224067021174. Accessed August 15, 2017.

RCN. (2009). https://www.forskningsradet.no/prognett-forny/Programme_description/1226485703385. Accessed August 20, 2017.

RCN. (2010). Midway evaluation of the centres for research-based innovation. Evaluation Division for Innovation.

RCN. (2011). *Norway's technology transfer offices: An eye for commercializing research results*. Available at https://www.forskningsradet.no/en/Newsarticle/An_eye_for_commercialising_research_results/1253969972866.

RCN. (2012a). https://www.forskningsradet.no/prognett-vri/Nyheter/Funding_instruments_for_regional_RD_and_innovation_evaluated/1253982002802. Accessed August 29, 2017.

RCN. (2012b). https://www.forskningsradet.no/prognett-energix/Programme_description/1253980140102. Accessed August 28, 2017.

RCN. (2013). Midterm evaluation of centres of environment-friendly energy research. Evaluation Division for Energy, Resources and the Environment.

RCN. (2014). *Revised work programme for the programme for user-driven research-based innovation (BIA)*. Oslo: Research Council of Norway.

RCN. (2015a). *Basic and long-term research within engineering science in Norway* (Report from the Principal Evaluation Committee). Oslo: Research Council of Norway.

RCN. (2015b). *Research for innovation and sustainability* (Strategy for the Research Council of Norway 2015–2020). Oslo: Research Council of Norway.

RCN. (2016a). *Three of five Norwegian EFC grants go to centres of excellence*. Available at https://www.forskningsradet.no/en/Newsarticle/Three_of_five_Norwegian_ERC_grants_go_to_Centres_of_Excellence/1254021775768.

RCN. (2016b). *Norway to fund eight new centres for environment-friendly energy research (FME)*. Available at https://www.forskningsradet.no/prognett-energisenter/Nyheter/Norway_to_fund_eight_new_Centres_for_Environmentfriendly_Energy_Research_FME/1254018487675.

RCN. (2017a). https://www.forskningsradet.no/prognett-energisenter/About_the_centres/1222932140914. Accessed August 14, 2017.

RCN. (2017b). https://www.forskningsradet.no/prognett-petromaks2/Programme_description/1253980921389. Accessed August 28, 2017.

Rotefoss, B., Pedersen, E., Jenssen, S. A., & Kolvereid, L. (2010). The case of Norway. In A. Lundström (Ed.), *Towards an entrepreneurship policy—A nordic perspective*. Stockholm: Swedish Foundation for Entrepreneurship.

Simanovska, J. (2013). *Green public procurement in Norway*. Available on http://lpmc.lv/uploads/media/Green_public_procurement_in_Norway.pdf.

Skattefunn. (2014). https://www.skattefunn.no/prognett-skattefunn/Funding_Opportunities_and_Eligibility/1254001716647?lang=en. Accessed August 20, 2017.

SIVA. (2017a). https://siva.no/norsk-katapult/. Accessed August 28, 2017.

SIVA. (2017b). https://siva.no/om-oss/?lang=en. Accessed August 15, 2017.

Smith, K., Dietrichs, E., & Nås, S. O. (1996). *The Norwegian national innovation system: A preliminary overview and assessment*. Available at http://www.oecd.org/sti/inno/2373824.pdf.

The World Bank. (2017). https://data.worldbank.org/. Accessed December 17, 2017.

Ville, S., & Wicken, O. (2013). The dynamics of resource-based economic development: Evidence from Australia and Norway. *Industrial and Corporate Change, 22*(5), 1341–1371.

Chapter 4
Case Study: Singapore

Abstract Singapore is a rapidly developing country with a diverse population sitting on one of the most strategic locations in the world—Strait of Malacca, which transports more than 50% of world's commercial goods. Over the past fifty years, the country has transformed itself from a technology user to a technology developer. The main source of human and financial capital in the country is the foreign talent and direct investment. The government has established various agencies and institutions, such as Agency for Science, Technology and Research, to propel domestic innovation in collaboration with MNCs. Most importantly, the government has maintained a liberal immigration policy to attract the overseas talent. Moreover, the IP Act and Competition Act balance each other to promote competition between low-tech and high-tech enterprises but at the same time these regulations ensure a level playing field for all actors in the innovation system of Singapore. Similar to other innovation-driven countries, Singapore has also developed public research institutions, clusters, and science parks. In parallel to the institutional and infrastructure development, there has been an equal emphasis on training and development of human capital. On other hand, a host of public-private partnership programs, such as Technopreneurship21 and GET-UP, ensures collaboration between industries, universities and government at all levels.

4.1 Background

Singapore is a small and emerging nation with its economy mainly based on foreign direct investments (FDIs). Since its political independence in 1965, Singapore went through several phases of industrial and technological upgrades and has established itself as one of the strongest and diversified economies in the world. The population of the country was 5.607 million in 2016 and was ranked as the third densest country in the world (The World Bank 2017). Singapore's GDP per capita (in Purchasing Power Parity) in 2016 was 88,003 International Dollars, fourth highest in the world (The World Bank 2017). Government's role has remained central in the technological

R. Pradhananga co-authored this chapter.

© Springer Nature Switzerland AG 2020   53
W. Nawaz and M. Koç, *Industry, University and Government Partnerships*
for the Sustainable Development of Knowledge-Based Society,
Management and Industrial Engineering, https://doi.org/10.1007/978-3-030-26799-5_4

development of the country, and R&D gained increasing attention in the late 1980s. Singapore's R&D expenditure in 2015 was 2.4% of the GDP, and the ratio of business and public sector expenditure was 3:2 (Agency for Science Technology and Research ASTAR 2015a). Singapore has organically developed a strong education and research sector and has created an environment that facilitates collaborations and knowledge sharing within and outside the country. As a result, Singapore established itself as a global innovation leader; GII of Singapore in 2017 is 58.69, the seventh highest among the 127 countries across the world (Cornell University, INSEAD, and WIPO 2017).

4.2   Institutional and Cultural Setting

The national innovation system of Singapore can simply be characterized by foreign capital, technology, and talent managed diligently by government (Wong 2015). The ability of the Singapore government to timely identify the potential market trends and deploy long-term strategic development plans has significantly benefited the economic, social and human capital development of the country. Singapore's economy until 1980 was largely labor-intensive and dependent on technology transfer from MNCs. National innovation system during this period focused on the development of infrastructure and human resources to absorb and exploit the new technologies rapidly (OECD 2013). The economic growth from 1980 to 1990 was technology and service intensive which shifted the economic focus of the country to innovation-driven R&D. Establishment of National Science and Technology Board (NSTB) and the release of the first National Technology Plan in 1991 were some early efforts to promote R&D in Singapore. Research collaborations increased significantly with the introduction of Research and Development Assistance Scheme (RDAS). For the past two decades, there has been a significant growth in the knowledge and innovation intensive services and manufacturing in Singapore. The sectors which remained dynamic throughout the organic development are electronics, IT, and biomedicine. The government has supported these sectors with innovative plans to emphasize on entrepreneurship and the long-term R&D.

In 2001, Ministry of Trade and Industry (MTI) established the Economic Review Committee to review Singapore's economic position and identify areas to strengthen national entrepreneurship and innovation. In the same year, NSTB was restructured as Agency for Science, Technology and Research (ASTAR). At present, there are three statutory boards which have the mandate to develop and implement science, technology and innovation policies in Singapore: MTI; Ministry of Education (MOE); and the National Research Foundation (NRF) established in 2006 as a secretariat to Research, Innovation and Enterprise Council (RIEC) which was chaired by the Prime Minister. These institutions are flexible enough to perform rapid upgrade of the policies which makes a significant impact on Singapore's innovation performance (Wong 2003).

Private sector, particularly the foreign firms, account for a large share of R&D activities and expenditure in Singapore. Local firms that engage in R&D activities are the technically advanced SMEs operating in various supporting industries. Other firms involved in R&D are either the state-controlled enterprises established by the Singapore government or the high-tech entrepreneurial start-ups which are usually founded by the university professors, researchers at public research institutes, and university students. Public sector R&D is conducted by universities and public research institutions. Until 1990, universities were the main actors for public sector R&D. With the establishment of public research institutions, the share of R&D in universities decreased. However, with the greater emphasis placed on basic R&D more recently, R&D in universities is again showing an increasing trend. Collaborations between private sector, mainly MNCs and the local firms, started as early as 1980s (Wong 1999a). However, the collaborations between universities/public research institutions and the private firms started only after early 2000s when a greater emphasis was placed on the commercialization of technologies by the government (OECD 2013).

One of the distinguishing features of Singapore's innovation system is its constant effort in the human development. The vocational training institutes and the Polytechnique institutes together with government and partner MNCs offered specialized training programs (Wong 2003). In addition, the curriculum of the universities is updated in consultation with the industry and international experts. The frequency of curriculum update is also relatively high in order to address the emerging needs of the country. To supplement the highly skilled professionals, the government has introduced initiatives that emphasize on the role of public research institutions in the training programs. Most importantly, the government has maintained a liberal immigration policy to attract the overseas talent.

4.3 Legislations/Regulations

In contrast to Norway, Singapore's legislative efforts are more towards the privatization side than nationalization. Singapore started a tax deduction scheme to promote R&D as early as 1980s (Wong 2003). Over the time, the R&D Tax Allowance Scheme under Tax Act was revised and different tax support programs were announced to promote R&D activities within and outside Singapore. Besides the generous tax incentives, Singapore has a strong IP governance system which attracts companies to conduct more innovative activities within the country. IP regulations, such as the Patents Act, Copyright Act, and Trademark Act, provide strong market power to the inventors, creators, and IP owners (Singapore Statutes Online SSO 2017). Moreover, generous financial incentives have been directed to encourage development and registration of IPs. The Singapore Patents Act, enacted in 1995, offers a 20-year right to the patent holder for making, using, or selling the patented inventions.

While the IP act provides some degree of market power to the creators or developers, the Competition Act constraints and balances the use of market power. First in place in 2004, the Competition Act prohibits three types of anti-competitive practices from private entities (Competition Commission Singapore CCS 2017). First, it restricts the agreements, decisions, and practices which prevent competition in the market. Secondly, it prohibits the abuse of dominant position in the market. Lastly, it prevents mergers that substantially lessen the competition. The act is applicable to all private sectors regardless of the ownership of the enterprise (national or international). Hence, the IP act and the competition act go hand in hand to promote competition but at the same time, these regulations ensure a level playing field for all actors in the innovation system of Singapore.

4.4 Intermediary Structures

Following are the key intermediary structures that facilitated IUGPs in Singapore:

4.4.1 Government/Semi-Government Research Institutions

Public research institutions in Singapore can be categorized into two groups: (i) the pioneer institutions under the Public Research Institutes, Centers and Consortia (PRICs); and (ii) several R&D centers created more recently under the Campus for Research Excellence and Technological Enterprise (CREATE), Research Centre of Excellence (RCE), and Center of Innovation (COI).

4.4.1.1 PRIC

With a mission to create a strong base of science and technology, Singapore government accelerated support for establishment of PRICs in late 1980s (Wong 2003). Molecular and Cell Biology institute, launched in 1985, is the first public research institute in Singapore. Initially, the PRICs focused on applied R&D due to the increasing technological needs of industries. Their involvement in fundamental R&D increased between 1995 and 2005. Until 2000, the PRIC enterprise collaboration was weak; the public research consortia were established to promote such collaborations between public and private sectors and to coordinate the practical translational of research (ASTAR 2017a). Most PRICs are housed within the universities and have academic staff of the hosting university as the principle investigators (OECD 2013). Recently, some PRICs have also established themselves as autonomous bodies. A majority of the PRICs are the institutes, centers and consortiums established under ASTAR. As of 2017, there were 21 such institutions managed by the Biomedical Research Council (BMRC) and Science and Engineering Research Council (SERC) of ASTAR. The R&D activities of these institutions cover a wide range of disciplines including biotechnology, medical technology, pharmacy, communications, chemicals, computational sciences, and manufacturing.

4.4.1.2 CREATE

CREATE, operational since 2006, is an international collaboration between research centers set up by top global universities, polytechnics, research institutes, and industries in Singapore (CREATE 2017). The centers under CREATE are recognized for establishing spinoffs and startups. The centers focus on four interdisciplinary areas of research: human systems, energy systems, environmental systems, and urban systems. As of 2017, there are 10 research centers under CREATE. Singapore-MIT Alliance for Research and Technology (SMART) Centre was the first to operate under the flagship of CREATE. SMART was built in collaboration between MIT, National University of Singapore (NUS), and Nanyang Technological University (NTU). CREATE has already established itself as a leading global research hub. In 2013, CREATE won the title of 'Laboratory of the Year' from the US-based R&D Magazine (NRF 2017a).

4.4.1.3 RCE

The RCE is a joint program hosted by NRF and MOE to strengthen research excellence in the universities in Singapore. Starting 2007, the program established 5 research centers dedicated to earth science, quantum technology, cancer science, mechanobiology and environmental life science. Each RCE center is hosted by a local university to conduct outstanding research in the focused area (NRF 2017a).

4.4.1.4 COI

COI is a program initiated by the Standards, Productivity and Innovation Board (SRING) of Singapore in 2007. The program aims to assist SMEs to upgrade their technological and innovative capabilities (SPRING Singapore 2017a). The COIs are established by SPRING Singapore in partnerships with industries and selected polytechnics and research institutes (Wong et al. 2010). The centers provide laboratory facilities, training courses, testing services, and technology consulting to SMEs. As of 2017, there were 8 COIs specialized for SMEs in electronics, environment and water, food, marine and offshore, materials, precision engineering, health products and supply chain management sectors.

4.4.2   Government Initiated Innovation Clusters

The small size of the country and stable political environment makes Singapore ideal to benefit from clusters. In this regard, the Singapore government has exercised three national level innovation clustering projects namely Biopolis, Fushionopolis, and the Mediapolis.

4.4.2.1   Biopolis

Launched in 2003, Biopolis is a biomedical research and technology hub in Singapore (Wong et al. 2010). It is a collocation of biomedical research institutes and companies in Singapore designed to foster collaborations and co-innovations in the sector. The integrated research network in Biopolis facilitates companies seeking improvement in R&D productivity and innovative biomedical solutions. As of 2017, Biopolis was a home to 4400 researchers from public and private sector in life sciences (Ascendas-Singbridge 2017). The cluster hosts more than 40 private companies including GlaxoSmithKline, Novartis and Ionis Pharmaceuticals, formerly known as Isis Pharmaceuticals. Biopolis is one of the most successful government efforts for innovation in Singapore, and it has significantly contributed to the rapid growth of biomedical science industry in the country (ASTAR 2013).

4.4.2.2   Fushionopolis

Fushionopolis is an integrated physical infrastructure dedicated to strengthening innovation in ICTs, physical sciences, and engineering industries (Ascendas-Singbridge 2017). It was launched in 2008 and since then it has housed several research institutes and corporate labs (OECD 2013). Fushionopolis promotes scientific research and technological breakthroughs through close cooperation and collaboration between public and private sectors.

4.4.2.3   Mediapolis

Following the success of Biopolis and Fushionopolis, Singapore government announced another cluster project, Mediapolis, dedicated to ICTs and media industry (Jurong Town Corporation JTC 2017). The infrastructure development is scheduled to be completed by 2020. Mediapolis aims to support creation of a new generation of media firms through joint effort between science and industry community (Ministry of Communication and Information, MCI Singapore 2012).

4.4.3 Science/Technology/Business Parks

Development of science parks in Singapore is a part of government policy to attract global MNCs to locate and invest in Singapore (Koh et al. 2005). The first Singapore Science Park was developed in 1980. The physical facilities of the park in the proximity of the national universities provided a perfect environment to promote innovation and technological development. In parallel, the government implemented other supportive strategies to ensure the presence of prominent MNCs in the park, such as tax incentives and creation of a network of potential domestic business actors. These efforts were extremely successful and the science park was fully occupied by mid 1990s (OECD 2013). Following the success of first science park, the development of second Singapore Science Park begin in 1993 (Ascendas-Singbridge 2017). Most of the tenants in the second science park were enterprises and research institutes related to ICT sector. The first and second Singapore Science Parks were both motivated to attract FDI, and clustering and coordination were less focused in both cases (Hu and Shin 2002; Koh et al. 2005). Therefore, in 2001, Singapore government came up with a plan to develop a large-scale integrated structure, One North, representative of science city or science district. The One North integrates existing science parks and offers a vibrant R&D environment with strong IP system. The focus of One North is to create informal networks and to facilitate greater knowledge exchange. As of 2017, One North was a home to 18,000 professionals from different public and private sectors (Ascendas-singbridge 2017).

4.4.4 Technology Transfer Offices

In 1992, NUS established NUS Industry Liaison, which was the first TTO in Singapore (Neubauer et al. 2013). The other autonomous universities of Singapore, Singapore Management University (SMU) and NTU followed suite and established office of research and tech-transfer and innovation and technology transfer office respectively. The technology transfer office is funded by the MOE and provides supports to other TTOs to foster close collaborations between the industry, university and research centers (Lim 2014).

4.4.5 Business Incubation Centers

Business incubators and accelerators in Singapore are growing steadily particularly from the last ten years. In addition, several programs, such as the NRF's Technology Incubator Scheme, are in place to support the startups and incubation process. For example, BLOCK71, also known as Blk71, located near the technology clusters Biopolis and Fushionopolis currently is a home to more than 30 incubators, accelerators and venture capitalists in Singapore and is operational since 2011 (Block71 2017). Blk71 is a joint effort between NUS Enterprise, SingTel Inov8 (the venture capital arm of the Singtel Group) and Media Development Authority of Singapore. Within a few years of its establishment, Blk71 established itself as an entrepreneurial hub which collocate start-ups, incubators and accelerators. It has been referred as "the world's most tightly packed Entrepreneurial ecosystem" by the Economist in 2014. Considering the success of Blk71, the Singapore government has announced development of two more physical infrastructures Blk79 and Blk73 to house more start-ups, incubators and accelerators.

4.5 Support Programs

Singapore's public-private sectors have initiated several support programs to promote collaborations between industry, academia and government as discussed in the following subsections.

4.5.1 Public Procurement

Public sector in Singapore has been energetic in using new technologies, especially IT (Wong 2003). By early 2000s, Ministry of Environment, Ministry of Communication and Information (MCI), and Singapore Telecommunications Limited were using the latest technologies in their fields. The share of government procurement from SMEs were also impressive. More recently, the government is planning to encourage use of crowdsourcing in public procurement to accelerate innovative solutions (MCI Singapore 2017). In addition, the government will encourage spiral contracting for the public agencies, which will allow the project to be awarded in phases and thus, will reduce the risk of loss.

4.5.2 Tax Incentives

Tax incentive is a primary strategy used by the Singapore government to attract FDIs in Singapore. In 1986, the government reduced the corporate income tax rate from 40 to 33%, which was one of the major tax cuts in the history of the country (Economic Review Committee 2003). The rate was subsequently lowered in the following years and from 2010, it has been fixed at 17%. The first announcement on R&D tax incentive in Singapore was made in 1980 which was applicable for the manufacturing sector only (Wong 2003). By 1990s, the government broadened the R&D tax incentive schemes to service sector as well. More elaborated R&D tax allowance schemes, such as 'R&D Tax Allowance Scheme (RDA)' and 'R&D

Incentive Scheme for Start-up Enterprise (RISE)', were introduced in 2008. The present R&D tax allowance scheme in Singapore includes four R&D tax deductions. Firstly, 'the basic tax deduction for R&D', continuing since 2008 (Inland Revenue Authority of Singapore IRAS 2015), offers 100% tax deduction for all qualifying R&D expenditures inside and outside of Singapore (Kaya and Bozdoğanoğlu 2016). Secondly, the 'productivity and Innovation Credit (PIC)' scheme, started in 2010, replaced the previous tax deduction schemes RDA and RISE. PIC offers additional 250–300% tax deduction on up to 400,000 Singapore Dollar (SGD) expenditure per year or on SGD 600,000 for qualifying SMEs (valid until 2018). The other tax deduction schemes in place since 2014 include 'enhanced tax deduction for R&D' and 'super deduction for R&D'. The former provides additional 50% deduction for R&D carried out in Singapore and the latter offers additional 50–100% tax deduction for R&D projects approved by the Economic Development Board (EDB). R&D expenditures that have received PIC do not qualify for the super deduction.

4.5.3 Internship, Training, Entrepreneurship and Innovation in Curriculum

Innovation in Singapore is structured on a solid foundation of entrepreneurship and training programs. In this regard, NTU's Technopreneurship and Innovation Program, started in 2002, appears to be the earliest training programs in the country. Although started late, Singapore's entrepreneurship system has been supported by several other programs, which helps the country to be competitive in the global innovation market. Additionally, Singapore has technical institutions, such as Institute of Technical Education (ITE), that are providing competitive training programs in coordination with national and international industries and businesses in the country. Universities are also actively engaging with industry leaders to offer industry-relevant curriculum and internship programs (Lim 2014).

4.5.4 Public-Private Partnership Programs

Singapore has a long history of strong collaborations between industry and training institutions. However, the collaborations between industries and universities/PRICs were less developed until late-1990s (Wong 1999b, 2003). The collaborations have remarkably increased in the last decade with significant efforts made by the government to promote entrepreneurship and R&D. In 2009, the government commenced Research, Innovation and Enterprise (RIE) to develop strategies to foster public-private R&D efforts. Some major initiatives undertaken by RIE include: Technopreneurship21, Growing Enterprises with Technology Upgrade (Get-up), Corporate Laboratory Initiatives, National Innovation Challenge (NIC), Thematic R&D

programs, Test Bedding and Demonstration of Innovation Research, ASTAR Collaborative Commerce Marketplace (ACCM), Collaborative Industry Projects (CIP), Partnerships for Capability Transformation (PACT), Technology Adoption Program (TAP), and Start-up/Gap Funds. ASTAR (2011) reported outstanding R&D performance in Singapore in terms of human capital, intellectual capital and industrial capital.

4.5.4.1 Technopreneurship21

Technopreneurship21 is a joint effort between NSTB and other government, private and educational agencies with an aim to nurture and build high-tech enterprises. The program was launched in 1999 and continued until 2005 (Wong 2015). At the beginning, the focus was on technology start-ups, however, later in early 2000s, the focus shifted to broader entrepreneurship in general. To achieve the objectives, the program adopted a two-prong approach: firstly, the investment and incentive schemes were developed to create favorable business environment and secondly technology and innovation were promoted in education sector.

4.5.4.2 GET-UP

GET-UP, established in 2003, is a multi-agency effort of ASTAR, EDB, SPRING Singapore and International Enterprise (IE) Singapore (Ho et al. 2015). The program uses existing financial assistance schemes and technical capabilities of ASTAR-PRICs to assist technology upgrading of SMEs in Singapore. It adopts an integrated approach under three schemes (2017b): Technology for Enterprise Capability Upgrading (T-Up); operation and technology road mapping; and technical advisory support. T-UP is the key scheme of GET-UP program. Under this scheme, PRICs and the local enterprises discuss and structure innovative project with potential benefit to the enterprises. Experienced researchers, scientists, and engineers from the PRICs are assigned to the work with the enterprises for a period of two years in order to upgrade the R&D capabilities of the enterprises and to create technologies with significant economic value. By 2011, the GET-UP program deployed 296 research scientists and engineers to 183 SMEs. Furthermore, the program helped in developing 161 roadmaps for 137 SMEs. A survey conducted by NUS Entrepreneurship Centre in 2009 revealed that the enterprises involved in GET-UP program projected their revenue to be twice in comparison to the ones which did not participate in the program (ASTAR 2011).

4.5.4.3 Corporate Laboratory Initiatives

ASTAR's Lab-in-RI, launched since 2004, is a pioneer program related to corporate laboratories in Singapore. The initiative, originally designed to support enterprises through R&D infrastructures available at public research institutes, was successful in developing strong public-private collaborations. As a result, several MNCs collaborated with PRICs to establish corporate R&D labs (ASTAR 2011). In 2013, NRF started another initiative, Corporate Laboratories in Universities, which supports foreign and domestic industries to establish laboratories in universities in Singapore (NRF, 2017a). The goal was to conduct industry related research in the academic environment while keeping industry as an active member of the research project (Lim 2014). As of 2017, 9 such laboratories were established, covering a wide range of areas such as environment, engineering, manufacturing and electronics.

4.5.4.4 NIC

NIC is NRF's national program to strengthen multi-disciplinary R&D in Singapore with active IUGP. The aim of the program is to foster the development of cutting-edge technologies of national significance in Singapore. Currently, three sector specific programs, i.e., Energy NIC, Land and Liveability NIC (L2 NIC), and NIC on Active and Confident Ageing, are operational (NRF 2017a). First in place in 2011, the Energy NIC seeks to achieve national level breakthroughs in energy efficiency and alternatives and carbon emission reduction. L2 NIC, launched in 2012, is a multi-agency funding awarded to innovative projects with potential of increasing Singapore's land capacity. NIC Active and Confident Ageing program, established in 2015, provides funding to promote innovative ideas and conduct research that can change the ageing experience in Singapore.

4.5.4.5 Thematic R&D Programs

NRF has launched dedicated programs, National Cybersecurity R&D program (NCR), Marine Science R&D Program (MSRDP), and Artificial Intelligence R&D program of Singapore (AI.SG) to support collaborative R&D projects in cybersecurity, marine science, and artificial intelligence sectors respectively. Launched in 2013, NCR is a joint program between NRF, National Security Coordination Centre, Cyber Security Agency, Ministry of Defense, Ministry of Home Affair, Infocommunications Development Authority of Singapore, and EDB. The aim of the program is to strengthen security, reliability, resiliency and usability of cyber infrastructures (NRF 2017a) and to create global market opportunities in cybersecurity through active collaboration between government agencies, academia, research institutes, and industries. MSRDP, announced in 2015 and hosted by NUS, funds training and collaborative R&D projects in marine science sector for a period of three to five years (MSRDP 2017). The goal of MSRDP is to promote environmental and marine sustainability among industries. In 2016, MSRDP grant was awarded to 7 competitive projects (NRF 2017a). AI.SG, launched in 2017, is a multi-agency program between NRF, Smart Nation and Digital Government Office, EDB, the Info-Communications Media Development Authority, SGInnovate and Integrated Health Information Systems operate the program (NRF 2017a). AI.SG funds AI-based R&D programs with active involvement of AI-based research institutions, start-ups, and companies in Singapore. The aim of the program is to strengthen Singapore's digital economy with innovation and wide application of Artificial Intelligence (AI).

4.5.4.6 Test Bedding and Demonstration of Innovation Research

The Test-Bedding and Demonstration of Innovation Research is NRF's funding initiative to facilitate deployment of technologies developed in public sector laboratories and institutions (NRF 2017a). In partnership with the local industries and public agencies, the initiative funds and provides a platform for testing and adoption of technologies with high potential to enhance service delivery. From 2015, the funding has been awarded to 6 technology adoption projects.

4.5.4.7 ACCM

ACCM is a portal launched by ASTAR in 2016 (ACCM 2017) to publish a list of company profiles in Singapore to help SMEs, MNCs, and academic and research institutions to identify their potential collaborators. SMEs and MNCs can form business collaborations by matching the technological needs of the companies while universities and research institutions can form research collaborations with the industries to support technological upgrade.

4.5.4.8 CIP

SPRING Singapore, in partnership with Trade Associations and Chambers, COI, and Solution Providers Singapore, established the CIP initiative in 2013. The objective of the program is to strengthen innovation and productivity capacity of SMEs in Singapore through development of solutions to the industry-specific challenges (SRPING Singapore 2015). The funding is awarded to competitive projects from consortia consisting of at least three SMEs. Approved projects are eligible for 70% funding for development and adoption of the solutions (ASTAR 2017c).

4.5.4.9 PACT

enlargethispage-24ptThe PACT program was launched by EDB in 2010 (SPRING Singapore 2013) to strengthen technological and innovative capabilities of SMEs in manufacturing sector. The program was expanded in 2013 to include all industrial sectors. After 2013, the program was administered by EDB and SPRING Singapore. PACT provides funding for collaborative projects between MNCs and local SMEs to facilitate knowledge transfer, capability upgrade, and development of test-bedding of innovative solutions. In 2017, a government-based PACT program was announced where government acts as a large organization and works together with SMEs/startups (SPRING Singapore 2017b).

4.5.4.10   TAP

ASTAR together with SPRING Singapore announced TAP program in 2013 to help local companies to get easy access to technology enhancements (ASTAR 2015b). The program supports collaboration between technology providers, such as PRICs, universities, and private technology integrators. The aim of TAP is to identify and translate new technologies into ready-to-go solutions that are easily adaptable by the local businesses; technology adaption projects in Singapore with productivity gain of at least 20% are eligible to apply for the TAP program. As of 2015, the program supported 1300 projects which benefited more than 900 companies.

4.5.4.11   Start-up/Gap Funds

Government agencies in Singapore offer various funding schemes to assist business start-ups. For example, Start-up Enterprise Development Scheme (SEED), originally launched by EDB in 2001 (EDB 2017) and later renamed to SRPING SEED in 2007, provides equity financing for local startups that have an innovative idea or product. The program is administered by the Startup SG Equity (a subsidiary of Startup SG) which is an umbrella organization established by the Singapore government to unify all government startup efforts in Singapore. By 2008, SEED funded more than 150 start-ups. The Business Angel Scheme (BAS) administered by SPRING Singapore was established in 2005 (Wong 2011). The aim of BAS is to promote proactive angel investments in innovative ideas. Early Stage Venture fund (ESVF) and Technology Enterprise Commercialization Scheme (TECS) are the other important start-up schemes offered by the Singapore government. ESVF is administered by NRF and was launched in 2008. The scheme offers investment in Singapore-based early stage high-tech companies. Similarly, SRING Singapore and ASTAR launched TECS in 2008 to provide funding to support innovative technological ideas at critical pre-market stage (ASTAR 2011). By 2013, the scheme supported more than 150 projects. Other gap funding schemes supporting commercialization of research include the Commercialization of Technology (COT) fund, flagship funding, and Central Gap fund. ASTAR's COT funding, started in 2006, supports development of working prototypes or functional processes adoptable by the market. More recently, NRF has introduced the Central Gap funding program at national level to support prototyping of translational research. The program funds translational research of significant commercial potential for a period of two years with a possible extension of one year (NRF 2017b). Also, ASTAR funds larger translational projects with significant commercial potential under ASTAR's flagship programs (ASTAR 2011), such as the Exploit Technology Flagship program, started in 2006, and the Translational and Clinical Research program, introduced in 2007.

Table 4.1 Summary of the IUGP trends and drivers in Singapore

	1960 and before	1960–1970	1970–1980	1980–1990	1990–2000	2000–2010	2010 and after
Governance		EDB, MOE	MTI		NSTB, SPRING Singapore/MTI	Economic Review Committee/MTI, A*STAR/MTI, NRF/RIEC	
Legislations				R&D Tax deduction Scheme under Tax Act, Copyright Act	Patents Act, Trademarks Act	Competition Act	
Intermediaries							
Research Institutes/Centers/ Consortiums				PRIC		RCE, COI	CREATE
Clusters						Biopolis, Fushionopolis	Mediapolis
Science Parks/Business Incubators/TTOs				Science Park	TTO		Blk71, Blk79, Blk73
Policies							
Public Procurement						Ministry of Environment, MCI, Singapore Telecoms using Latest Technologies	Crowdsourcing, Spiral Contracting

(continued)

Table 4.1 (continued)

	1960 and before	1960–1970	1970–1980	1980–1990	1990–2000	2000–2010	2010 and after
Tax Incentive				R&D Tax Incentive for Manufacturing Sector	R&D Tax Incentive for Service Sector	RDA, RISE, Basic Tax Deduction for R&D	PIC, The Enhanced Tax Deduction for R&D, Super Deduction for R&D
Academic Entrepreneurship and Innovation Programs						Academic Technopreneurship and Innovation Program	
Partnership Programs				RDAS	Technopreneurship21	Get-UP, Lab-in-RI, SEED, BAS, ESVF, TECS, COT, Exploit Technology Flagship Program, Translational and Clinical Research	Corporate Laboratory@University, Energy NIC, L2 NIC, NIC Active and Confident Ageing Program, NCR, MSRDP, AI.SG, Test Bedding and Demonstration of Innovation Research, ACCM, CIP, PACT, TAP, Central Gap Fund

A summary of the chapter has been tabulated in Table 4.1.

References

ACCM. (2017). https://accm.a-star.edu.sg. Accessed September 12, 2017.

Ascendas-Singbridge. (2017). http://www.ascendas-singbridge.com/en/our-properties/singapore. Accessed September 12, 2017.

ASTAR. (2011). *STEP 2015: Science, technology & enterprise plan 2015*. Singapore: Planning and Policy Department, ASTAR.

ASTAR. (2013). *Singapore's biopolis: A success story*. Available at https://www.a-star.edu.sg/News-and-Events/News/Press-releases/ID/1893/Singapores-Biopolis-A-Success-Story.aspx.

ASTAR. (2015a). *National Survey of Research and Development Singapore 2015*. ASTAR December 2016.

ASTAR. (2015b). *A*STAR deepens technology adoption for SMEs through new initiatives*. ASTAR, April 16, 2015.

ASTAR. (2017a). https://www.a-star.edu.sg/About-A-STAR/Research-Entities/Overview.aspx. Accessed September 12, 2017.

ASTAR. (2017b). https://www.a-star.edu.sg/simtech/industry/technology-upgrade-get-up.aspx. Accessed September 4, 2017.

ASTAR. (2017c). https://www.a-star.edu.sg/Collaborate/Programmes-for-SMEs/Forging-Partnerships.aspx. Accessed September 12, 2017.

Block71. (2017). http://www.blk71.com/. Accessed September 12, 2017.

CCS. (2017). https://www.ccs.gov.sg/legislation/competition-act. Accessed September 11, 2017.

Cornell University, INSEAD, WIPO. (2017). *The Global Innovation Index 2017—Innovation Feeding the World*. Ithaca, Fontainebleau, and Geneva.

CREATE. (2017). https://www.create.edu.sg/about-create. Accessed September 12, 2017.

Economic Review Committee. (2003). Restructuring the tax system for growth and job creation. In *Report of the Economic Review Committee*.

EDB. (2017). https://www.edb.gov.sg/content/edb/en/about-edb/company-information/our-history.html. Accessed September 20, 2017.

Ho, Y.-P., Hang, C.-C., Ruan, Y., Wong, P.-K. (2015). Transferring knowledge from PRIs to SMEs via manpower secondment: The case of Singapore's GET-UP program. In *XIII Triple Helix International Conference*, Beijing, August 21–23, 2015.

Hu, A. G., & Shin, J.-S. (2002). Climbing the technology ladder. In A. T. Koh, K. L. Lim, W. T. Hui, B. Rao, & M. K. Chng (Eds.), *Singapore economy in the 21st century*. Singapore: McGraw Hill.

IRAS. (2015). *IRAS e-tax guide: Research and development tax measures*. IRAS.

JTC. (2017). http://www.jtc.gov.sg/industrial-land-and-space/Pages/mediapolis.aspx. Accessed September 11, 2017.

Kaya, I. H., & Bozdoğanoğlu, B. (2016). Research and Development (R&D) tax incentives in Singapore. *International Journal of Humanities and Management Sciences, 4*(2), 181–183.

Koh, F. C. C., Koh, W. T. H., & Tschang, F. T. (2005). An analytical framework for Science Parks and Technology Districts with an application to Singapore. *Journal of Business Venturing, 20*(2), 217–239.

Lim. (2014). Linkage and collaboration between universities and industries in Singapore. In *SEAMEO RIHED regional seminar on linkage and collaboration between the higher education institutions and industries*. https://doi.org/10.13140/2.1.3840.6402.

MCI Singapore. (2012). https://www.mci.gov.sg/pressroom/news-and-stories/pressroom/2012/5/mediapolis–onenorth?page=87. Accessed September 12, 2017.

MCI Singapore. (2017). https://www.gov.sg/microsites/budget2017/press-room/news/content/ how-can-government-procurement-support-innovation-and-growth. Accessed September 11, 2017.

MSRDP. (2017). http://www.msrdp.sg/. Accessed September 12, 2017.

Neubauer, D., Shin, J. C., & Hawkins, J. N. (2013). *The dynamics of higher education development in East Asia: Asian cultural heritage, western dominance, economic development, and globalization.* New York: Palgrave Macmillan.

NRF. (2017a). https://www.nrf.gov.sg/programmes. Accessed September 12, 2017.

NRF. (2017b). https://www.nrf.gov.sg/funding-grants. Accessed September 20, 2017.

OECD. (2013). *OECD reviews of innovation policy: Innovation in Southeast Asia.* Paris: OECD Publishing.

SPRING Singapore. (2013). https://www.gov.sg/resources/sgpc/media_releases/SPR%20SPORE/ press_release/P-20131031-3. Accessed September 20, 2017.

SPRING Singapore. (2015). *Factsheet for collaborative industry projects.* Posted online on 21 May 2015.

SPRING Singapore. (2017a). https://www.spring.gov.sg/Growing-Business/Grant/development-areas/Pages/TI-Centres-Of-Innovation.aspx. Accessed September 12, 2017.

SPRING Singapore. (2017b). *Enhanced Partnerships for Capability Transformation (PACT) to Drive Government Lead Demand (Gov-PACT): Factsheet.* SPRING Singapore, 3 March 2017.

SSO. (2017). http://statutes.agc.gov.sg. Accessed September 11, 2017.

The World Bank. (2017). https://data.worldbank.org/. Accessed December 17, 2017.

Wong, P.-K. (1999a). National innovation systems for rapid technological catch-up: An analytical framework and a comparative analysis of Korea, Taiwan and Singapore. In *DRUID Summer Conference on National Innovation Systems, Industrial Dynamics and Innovation Policy, Rebuild,* Denmark, June 9–12, 1999.

Wong, P.-K. (1999b). University-industry technological collaboration in Singapore: Emerging patterns and industry concerns. *International Journal of Technology Management, 18*(3/4), 270–285.

Wong, P.-K. (2003). From using to creating technology: The evolution of Singapore's national innovation system and the changing role of public policy. In S. Lall & S. Urata (Eds.), *Foreign direct investment, technology development and competitiveness in East Asia.* UK: Edward Elgar Publishing.

Wong, P.-K. (2011). *Overview of angel investing in Singapore.* Available at https://www.techinasia.com/overview-of-angel-investing-in-singapore.

Wong, P.-K. (2015). *Singapore's evolving national innovation system.* Available at http://www.adb-asianthinktanks.org/sites/all/themes/webmate-responsive-theme/knowledgeresources/Singapore%E2%80%99s%20Evolving%20National%20Innovation%20System_Wong.pdf.

Wong, P.-K., Ho, Y.-P., & Singh, A. (2010). Industrial cluster development and innovation in Singapore. In A. Kuchiki & M. Tsuji (Eds.), *From agglomeration to innovation.* London: Palgrave Macmillan.

Chapter 5
Case Study: Qatar

Abstract Qatar is entirely different from the benchmark countries in terms of its geography, climate, size, mix of population, culture and thus, economic conditions. It is located on the tiny branch out of the Arabian Peninsula in the eastern side of the Arabian Gulf. It is smaller than the size of state of Connecticut with around only 12,000 km^2 of pretty flat and sandy land. The country is considered quite young as it gained its independence in 1971, with small and homogenous native population (i.e., estimated to be around 300,000) as a minority among its quite internationalized total population of around 2.75 million as of 2018. It is almost entirely economically dependent on its abundant oil and gas reserves (i.e., third largest natural gas reserves in the world). The leadership of the country has taken many steps in the past two decades to attempt to transform the hydrocarbon-based economy to knowledge-based economy. The most prominent step among these is the introduction of Qatar National Vision 2030, which sets out the roadmap of economic transformation and human development in Qatar while protecting its physical, natural and cultural environment. In addition, Qatar Foundation (QF), an initiative of the Royal Family, has been phenomenal in Qatar's effort to develop a knowledge ecosystem. At the same time, enacting the law of 'Protection of Intellectual Property and Copyright', 'Establishing Free Zone', and 'Patents Law' has clearly demonstrated the intentions and determination of the leadership towards a knowledge-based sustainable development. Furthermore, under the umbrella of QF, Qatar is home to eight international branch campuses which help the country in exploiting its local talent on one hand and attracting the foreign talent on the other. Moreover, Qatar has two national universities and three national research institutions operating in proximity of the international branch campuses, science park, and incubation center. The location of knowledge-intensive institutions in close vicinity in the Education City, which is home to the QF and its member institutes and branch campuses, allow these institutions to collaborate and compete at the same time. Other notable institutions playing key roles in the national innovation system are Qatar Development Bank (QDB), Qatar National Research Fund (QNRF), Qatar Science and Technology Park (QSTP), and the recently established Qatar Research Development Innovation (QRDI) council which offer various necessary funding and support programs to lift the innovational and technological quotient of the country to international standards.

© Springer Nature Switzerland AG 2020   71
W. Nawaz and M. Koç, *Industry, University and Government Partnerships*
for the Sustainable Development of Knowledge-Based Society,
Management and Industrial Engineering, https://doi.org/10.1007/978-3-030-26799-5_5

5.1   Background

The state of Qatar, an oil and gas rich country, and a member of Gulf Cooperation Council (GCC), has come forth as one of the most promising emerging economies of the world. Since the discovery and subsequent exploration of oil and gas reserves in 1940s and 1950s, the economic system of Arabian Peninsula has been primarily driven by the hydrocarbon resources (Abduljawad 2013). Prior to the discovery of oil, the population of Qatar was around 30,000 (Elsheshtawy 2008). However, with the influx of oil revenue starting from 1950s, especially after 1990s, Qatar witnessed tremendous growth in terms of population as well, mainly due to expatriates which represent around 90% of the total population (Ministry of Development Planning and Statistics MDPS Qatar 2016). As of 2015, population of Qatar was 2.4 million, which was 0.6 million in 2000 and 0.3 million in 1986. Qatar's continuing economic growth through abundant resources of oil and gas has been reflected in its GDP; in 2016, Qatar had the highest GDP per capita (in Purchasing Power Parity) in the world, with 127,728 International Dollars per inhabitant (The World Bank 2017). Nevertheless, the government of Qatar realized that this growth pattern cannot be sustained in the long term through the use of natural resources. Therefore, in 2008, Qatar launched its strategic vision to transform its economy to knowledge-based economy. Over the last ten years, the country has hugely expanded its network of HEIs. As of 2017, the country has a network of 16 HEIs (Ministry of Education and Higher Education, MOE 2017), many of which are the international branch campuses (IBCs) of well-known universities. The R&D expenditure of Qatar in 2015 was 0.51% of its GDP, which according to the survey conducted by the Ministry of Development Planning and Statistics (MDPS) in collaboration with Qatar Foundation R&D was second largest expenditure among the GCC countries (MDPS Qatar 2017). The survey also revealed a rise of more than 50% R&D personnel in Qatar between 2012 and 2015. Despite these efforts, the 2017 GII of Qatar was 37.9, forty-ninth highest among the 127 countries across the world (Cornell University, INSEAD, and WIPO 2017).

5.2   Institutional and Cultural Setting

After its independence from the British rule in 1971, the first effort made by the country towards developing the knowledge infrastructure was the establishment of Qatar's first national College of Education in 1973. This was followed by the establishment of Qatar University (QU), first university in the country, in 1977. In 1980's, there was no notable development in terms of institutionalizing knowledge and research. Nevertheless, in 1995, it was the establishment of Qatar Foundation (QF), a non-profit organization, which turned out to be a game-changer for the country. Although QF is registered as a private organization, it's funding and necessary support comes from the state government since it was founded by the then-Emir (Monarch) Sheikh

Hamad bin Khalifa Al Thani and his wife Sheikha Moza bint Nasser (Abduljawad 2013).

The most important milestone for Qatar in transforming to a knowledge-based economy is the development plan of the country—Qatar National Vision (QNV) 2030. Launched in 2008, QNV is a robust bureaucratic framework which has been articulated at a visionary level. QNV's operational translation was carried out through Qatar National Development Strategy (QNDS) 2011–2016 in order to define the roadmap for realizing the optimistic goals of QNV. At the same time, a comprehensive gap analysis was carried out by Qatar Science and Technology Park (QSTP) to identify research needs of national significance. The analysis resulted into the Qatar National Research Strategy (QNRS) that has a mandate to align the R&D activities in Qatar with the objectives of QNV. QF, being the custodian of QNRS, has taken numerous initiatives to address the National Grand Challenges which include Water Security, Energy Security, Cyber Security, and Health. In addition, many institutions were established with an aim to boost the knowledge-based economy under the flagship of QF: Education City (EC) was launched in 1997 and was officially inaugurated in 2003 to host several international university branch campuses and a national university; Qatar National Research Fund (QNRF) was established in 2006 as the primary research-funding agency in the country with an aim to foster research culture in Qatar (QNRF QNRF 2017a); QSTP was inaugurated in 2009 with an aim to develop it as an innovation hub in the region; Qatar Computing Research Institute (QCRI) was established in 2010 to enhance the computing competence in the country through R&D in this area; Qatar Environmental & Energy Research Institute (QEERI) was launched in 2011 with the mandate of advancing research in energy and water sector; and Qatar Biomedical Research Institute (QBRI) was founded in 2012 with an aim to advance the healthcare system through innovation in prevention, diagnosis, and treatment. In addition to these subsidiaries and other joint ventures, QF has its own R&D arm, called Qatar Foundation Research & Development (QF R&D) department. The operational framework of QF R&D, as shown in Fig. 5.1, is based around the monitoring and evaluation of the research activities carried out by the research institutes in Qatar in order to ensure that these activities remain in line with the national research priorities.

QF-R&D has identified following seven crucial domains of research in its Integrated Five-Year Business Plan (2013–2018) which are to be targeted through collaborations between all members of QF (Merekhi 2013):

- Enterprise Capabilities (workforce development, research facilities, research management system)
- Energy and Environment (renewable energy, energy efficiency, environmental quality, sustainability and others)
- Computing and IT (humanitarian technologies, bioinformatics, cyber security, national computing infrastructure)
- Health and Life Sciences (Diagnosis, treatment and prevention, genomic, personalized healthcare, biobank)
- Social Science, Arts and Humanities (social well-being)

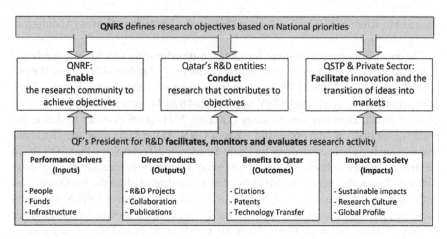

Fig. 5.1 The institutional framework of research entities and facilitators in Qatar (Bassil et al. 2017; Merekhi 2013)

- Entrepreneurship and Commercialization (IP and technology transfer)
- Global Engagement (global partnership, high profile communication of research results).

Nothing to take away from QU for the exceptional part it has played in establishing the plinth of the knowledge ecosystem and promoting innovation through research projects and university-based centers (Abduljawad 2013, 2015), but it is primarily QF which has developed a strong culture of creativity and research excellence within the Qatari society. The primary reason for this superiority is the presence of international research-driven university branch campuses in QF's EC. The consistent support offered by QF established a precedent of institutionalizing all new research and innovation centers under the flagship of QF. Not only that these units operate under the leadership of QF, these are physically located nearby QF headquarters and each other. Being in the proximity provides an added advantage to these institutes to collaborate and compete with each other in a positive working environment. As a result, QF has emerged as the spearhead of endeavors undertaken by the country to develop a culture of science, technology, research and innovation. QF has been remarkable in its efforts to support Qatar's mission to transform from hydrocarbon to knowledge-based economy. In addition to establishing the Education City (EC) to house multiple international branch campuses, such as Texas A&M College of Engineering, Georgetown School of Foreign Services, Carnegie Mellon Business School, Computer and Biological Sciences, Weill Cornell Medical School, VCU Art School, Northwestern Journalism and Communication School, QF also established another national graduate level university, Hamad bin Khalifa University (HBKU) in 2010. HBKU currently offers more than 20 Ph.D. and MS level programs under six different colleges – College of Science and Engineering, College of Health and Life Sciences, College of Humanities and Social Sciences, College of Law, College of Public Policy and College of Islamic Studies. QF has also led the efforts to establish a new national entity, Qatar Research Development and Innovation (QRDI) council, to coordinate all national R&D and innovation activities. The specific role, structure and impact of QRDI is yet to be seen.

Industries in Qatar are mainly in the oil and gas sector, which is the primary source of revenue to the national economy, i.e., 55% of GDP (OPEC 2016). Moreover, most of the enterprises in oil and gas sector, and industrial sector in general, are government owned. The negligible existence of the private sector makes it difficult to differentiate between the initiatives taken by the industries or government since, in most cases—directly or indirectly, the industries belong to the state. For example, Qatar Petroleum (QP) is a state-owned public corporation established since 1974 and controls the upstream and downstream of oil and gas sector in Qatar, which includes exploration, production, transportation, storage, marketing and sales. At the same time, other major industrial players in Qatar are either subsidiaries or joint ventures of QP, such as Qatar gas, Ras gas, Oryx GTL, Laffan Refinery, North Oil Company, Qatar Petrochemical Company (QAPCO), Qatar Steel, Qatalum, QChem, Qatar Fertiliser Company (QAFCO) and others.

Some industries in Qatar have their own research and development departments which are responsible for conducting applied research. For example, Qatar Steel has R&D department, which is established since 2012 with an aim to enhance the product innovation capabilities. On the fundamental R&D level, the industries collaborate with the research institutions and universities. Such partnerships can be in the form of projects (e.g., efficient use of treated sewage effluent in district cooling—a project of Hamad Bin Khalifa University, Qatar Cool, and Ministry of Municipality and Environment); joint ventures (e.g., Qatalum's Cast-house established a joint collaboration project with QU in 2015 to develop metallurgical competence in Qatar); centers (e.g., Mary Kay O'Connor Process Safety Center ext. Qatar—a center established in 2013 by the Mary Kay O'Connor Process Safety Center in US, Qatar Petroleum, and Texas A&M University at Qatar); and mutual support agreements (e.g., QP and QU's memorandum of understanding in 2012 to develop a working relationship between the two entities). The cooperation between the industry, university and research institutions boost the research and innovation competence in Qatar on one hand and develop and enhance a knowledge-intensive culture through mutual consultation, services, and training programs on the other. The number of such collaborations have significantly increased in last decade (Abduljawad 2013).

The limited private entities also support the national priorities in their way. For example, ConocoPhillips Qatar launched Global Water Sustainability Center in 2010 with an aim to offer innovative technological solutions for clean water production and management, and desalination in Qatar. Similarly, Shell operates a research and development center, Qatar Shell Research and Technology Centre (QSRTC), since 2008 with a promised funding of USD100 million over the span of ten years to offer innovative solution for the energy and water grand challenges of Qatar. Also, ExxonMobil's research center in Qatar, ExxonMobil Research Qatar, launched in 2009, conducts research in environmental management, water reuse, LNG safety and coastal geology. Other big names, such as Total, Microsoft, Siemens, Rolls-Royce, and General Electric have their research centers in QSTP.

On the governmental side, major governing bodies involved in the transformation of Qatar to a knowledge-based economy include MOE, Ministry of Economy and Commerce (MEC), Ministry of Transportation and Communication (MOTC), and the Ministry of Municipality and Environment (MME). However, direct involvement of the state government in the IUGPs is presumed limited. The apparently restricted involvement of government in IUGPs is mainly due to the indirect involvement of government in the knowledge infrastructure; the key players of IUGP in Qatar discussed thus far are either owned or funded or supported by the government. For example, Qatar Carbonates and Carbon Storage Research Centre (QCCSRC) is a ten-year strategic collaboration between QP, Shell, QSTP, and Imperial College London worth USD 70 million. Although the Qatari government does not appear on the face of this project, involvement of QP and QF is an evidence of government's support to this initiative. Another way of looking at it is through the board of directors of QP; in 2017, 3 out of 7 board directors of QP were government Ministers, including Minister of Energy and Industry, Minister of Economy and Commerce, and Minister of Finance. Hence, approval of investment in the collaborative R&D projects, such as QCCSRC, indirectly shows the support from the government.

The participation of government, however, becomes more visible when it comes to providing funding opportunities to Qatari nationals under especially designed programs; there are many scholarship opportunities provided by the government to Qatari nationals to pursue their education in the local or oversees universities. In addition to the government supported scholarships, industries and universities, particularly QP, QU and QF, also offer sponsorship and scholarship programs for supporting education and research.

5.3   Legislations/Regulations

In Qatar, there is no legal binding for any IUGP actor to be involved in such a partnership, however, the development of legislative framework over time, as shown in Table 5.1, is an evidence of country's commitment to transform to a knowledge-based economy.

5.4   Intermediary Structures

Over the past two decades, Qatar has established R&D departments, national research institutions, clusters, science park, innovation centers/consortiums, business incubation centers and TTOs to stimulate technological development through partnerships between academia, industries and government. However, as mentioned before, many of these entities operate under the flagship of QF.

Table 5.1 Qatar's legislative initiatives to transform into a knowledge-based economy

Law description	Law No. with date	Comments (if any)
Organizing scholarships	Law No. 9 of 1976	The purpose of granting scholarships was not only to facilitate the students in scientific, technical or practical studies but also to enable grantees to conduct research and attend practical training courses which could enhance their skillset
Establishment of QU	Law No. 2 of 1977	The law has been repealed by the law # 34 of 2004—Organization of Qatar University
Establishment of the center for scientific and applied research at QU	Emiri resolution No. 13 of 1980	–
Establishment of the center for educational research at QU	Emiri resolution No. 14 of 1980	–
Establishment of the documentation and humanities research center at QU	Emiri resolution No. 15 of 1980	–
Ratification of the educational, cultural and scientific agreement between the state of qatar and the syrian Arab Republic	Decree No. 30 of 1983	The Educational, cultural and scientific agreement between the state of Qatar and the Syrian Arab Republic was signed in Damascus on 14 June 1979
Protection of intellectual property and copyright	Law No. 25 of 1995	Repealed by Decree-Law No. 18 of 2009—abolishing certain Laws; however, the contents of the law were reflected in the Protection of Copyright and Neighboring Rights, Law No. 7 of 2002
Establishment of the Qatar Aeronautical College	Law No. 9 of 1996	The college was founded in 1977 as the civil aviation college for the GCC but later it changed its name to Qatar Aeronautical College and expanded its programs in 1996. However, the law has been repealed by Decree-Law No. 18 of 2009—Abolishing Certain Laws

(continued)

Table 5.1 (continued)

Law description	Law No. with date	Comments (if any)
Permission for citizens of GCC to practice economic activity in educational fields	Law No. 7 of 1997	It not only provided a provision for further investment in educational sector but also an opportunity of educational collaboration between the GCC countries
Establishing Qatar Technical College	Law No. 14 of 1998	The college was eventually abolished through Decree-Law No. 23 of 2002—Abolishing Qatar Technical College
Exempting Qatar University from certain provisions of Law No. 8 of 1976—organizing bids and tenders, and establishing the university bids committee	Council of Ministers Resolution No. 33 of 2001	The amendment was made to facilitate the educational institute in expediting the biding and tendering process. The same provision was also granted to the Ministry of Education and Higher Education through Council of Ministers Resolution No. 19 of 2000
Trademarks, trade indications, trade names, geographical indications and industrial designs and templates	Law No. 9 of 2002	–
Establishing the supreme education council	Decree-Law No. 37 of 2002	The council was merged into the Ministry of Education and Higher Education after the Emiri Resolution No. 9 of 2016 on the organizational structure of the Ministry of Education and Higher Education
Establishment of appreciative and incentivising state awards on sciences, arts and literature	Law No 11 of 2003	–
Protection of trade secrets	Law No. 5 of 2005	–
Protection of layout designs of integrated circuits	Law No. 6 of 2005	–

(continued)

Table 5.1 (continued)

Law description	Law No. with date	Comments (if any)
Establishing a free zone in QSTP	Law No. 36 of 2005	The aim of free zone was the promotion and support of scientific, applied and technological research. A Customs Department was later established at the main entrance of the Free Zone through Resolution No. 16 of 2011
Patents Law	Decree Law No 30 of 2006	The decree law was approved for the administration of IP and related laws, and to formulate the regulatory body for patents and technology transfer
Establishing the centre for creative leadership	Emiri resolution No. 22 of 2008	The original suggested name of the center was Qatar Leadership Centre which was later replaced with the name 'Centre for Creative Leadership' through Emiri Resolution No. 46 of 2010—Amending Certain Provisions of Emiri Resolution No. 22 of 2008
Approving the comprehensive vision for development "Qatar national vision 2030"	Emiri Decree No. 44 of 2008	–
Establishing the center for the protection of intellectual property rights	Emiri Decree No. 53 of 2009	This IP regulatory body was established under the Ministry of Justice. Later on, the law was repealed by Emiri Decree No. 20 of 2014 and a department of Intellectual Property Rights Protection was established under the MEC
Organization of the supreme education council	Emiri resolution No. 14 of 2009	–
Establishing Qatar development of small and medium-sized enterprises	Emiri resolution No. 17 of 2011	The Authority was established with an aim to encourage the establishment of enterprises and to support the existing enterprises in increasing their proportion of the contribution to the GDP

(continued)

Table 5.1   (continued)

Law description	Law No. with date	Comments (if any)
Educational voucher system	Law No. 7 of 2012	This law has been established to cover the educational cost of Qatari student by the following mechanism: • The law mandatories Governmental entities to pay the value of the educational vouchers for (a) the children of employees in these entities and (b) the children of those who are to be retired • The Supreme Education Council shall pay the value of the educational vouchers for the children of employees in non-governmental entities and the children of people who do not work in any of the entities

This is not a comprehensive list of relevant laws; some of the laws could not be accessed while the others were available only in Arabic language. A new law, under the MEC, to regulate 'Public Private Partnership' in Qatar will be issued soon

5.4.1   Government/Semi-government Research Institutions

Qatar has three national research institutes: QCRI, QEERI, and QBRI which are members of QF (QF 2016a) and are operational under the umbrella of Hamad Bin Khalifa University (HBKU)—the only national university in EC. The national research institutes are autonomous to collaborate with private businesses, universities, government agencies, innovation centers and parks, and international organizations. In addition, a national translational healthcare research institute, Interim Translational Research Institute (iTRI), has been established since 2014 under Hamad Medical Corporation (HMC). Other examples of national research and innovation centers include HMC's Medical Research Center (MRC), QF's Sidra Medical and Research Center (SMRC), Qatar Mobility Innovation Center (QMIC), and Qatar Bio Bank (QBB). In March 2017, the country formed a national level innovation-oriented consortium called Qatar Innovation Community (QIC) to support innovation in the country by facilitating the engagement between the academic community and entrepreneurs.

5.4.1.1 QCRI

The goal of QCRI is to tackle the large-scale computing challenges which are relevant to the national priorities of Qatar (QCRI 2017a). The core competencies of QCRI include Arabic language technologies, social computing, data analytics, distributed systems, cyber security and computational science and engineering. QCRI engages with international organizations, such as Boeing, World Bank, Microsoft, and Google, for carrying out cutting-edge research (QCRI 2017b). The institute recognize government agencies, such as Supreme Education Council and Qatar Statistic Authorities, as its stakeholders and involve them in research activities and projects to address the local needs. As one of the strategic objectives of QCRI is to have commercial impact on Qatar's economic diversification, the institute works in close collaboration with QMIC and QSTP to realize the most feasible commercial application of QCRI's research outcomes. From 2010 to 2014, QCRI had a total of 140 research staff, more than 350 peer reviewed publications, 35 software applications, 73 patents filed in the US and UK, 4 licensed technologies, and 1 startup (QCRI 2014).

5.4.1.2 QEERI

QEERI leads the national research in energy and water sector in Qatar (QEERI 2017a). The institute plays a vital role in developing IUGPs because of its research alignment with the industrial and governmental goals. Equipped with 14 labs, solar monitoring, and air quality monitoring station, QEERI provides state of the art facilities to the researchers with an aim to enhance sustainability and innovation in the country. On the domestic level, QEERI collaborates with private businesses (such as Qatar Solar Technologies and Green Gulf), government organizations (such as Qatar General Electricity & Water Corporation and Qatar Water and Electricity Company), and ministries and agencies (including Ministry of Environment, Ministry of Development, Planning and Statistics, and Supreme Committee for Delivery and Legacy) (QEERI 2017b). At the same time, QEERI join hands with national and international universities, including Texas A&M University at Qatar, QU, Drexel University-US, and Hanyang University-South Korea. QEERI's international partners include Swiss Supercomputing Center, Potsdam Institute for Climate Impact Research-Germany, and National Institute of Materials Science-Japan. Projects in QEERI are considered eligible for getting funded by the Qatar National Research Fund. In addition, the institute has its own Sponsored Research Office which oversees Sponsorships and Awards of the research projects. In the fiscal year 2014–2015, researchers and scientists at QEERI published 133 peer reviewed journal articles and 77 conference papers and presentations, and submitted 34 invention disclosures (QF 2015).

5.4.1.3   QBRI

The aim of QBRI is the transformation of healthcare in Qatar through science and innovation (QBRI 2017a). The institute comprise of three research centers: Cancer Center; Diabetes Center; and Neurological Disorders Center. As a disease-focused institute, QBRI plays a vital role in translation of medical research to practical outcomes. QBRI works in close collaboration with Riken Institute-Japan, QU, HMC, Weill Cornell Medical College in Qatar and Sidra Medical and Research Center (QBRI 2017b). Some of these partnerships extend beyond research projects to training and development of scientists and staff. The institute also organizes various public awareness lectures, conferences and workshops to highlight the importance of medical research in the region. From 2012 to 2016, QBRI published 72 peer reviewed research papers, and mentored three technology start-up companies in Qatar (QF 2016a). The number of publications of QBRI are lower than the other two national research institutes because of the lack of availability of research space for QBRI staff, up until 2015 (QF 2016a).

5.4.1.4   iTRI

iTRI supports healthcare research in the country to enable the translation of scientific discoveries to new medicines and applied technologies and treatments. Similar to the incubation and commercialization centers, iTRI provides state-of-the-art facilities including clinical studies unit with consultation rooms, wet laboratories for scientist, office space, and seminar and meeting rooms.

5.4.1.5   MRC

Established since 1998, MRC has a history of supporting medical research, which is one of the three pillars of HMC. The center aims to generate evidence based new knowledge that can be applied in every day medical practices. For this purpose, MRC manages not only the internal research activities but the externally funded projects as well. The research grants available at MRC are assessed on the basis of scientific merit. The outcomes of the research are made publically available, in most case. In 2016, the center published more than 100 research papers. The best research outcome at the center receive awards and recognitions every year. The center also works as a mentoring institute for the medical researchers to ensure that the researchers opt Good Clinical Practices in conducting research.

5.4.1.6   Sidra Medical Research Center (SMRC)

SMRC is a QF project, worth USD 7.9 billion, initiated with an aim to address three national goals: (1) world-class patient care; (2) medical education; and (3) biomedical research. Several scientific groups have been established in the center, including genetic medicine, immunology, bioinformatics and others, which operate under three research divisions including experimental genetics and genomics, translational medicine, and biomedical informatics. The major portion of research funding at SMRC comes from the QNRF grants; an estimated forecast of USD 42,315,000 through QNRF between 2015 and 2020 (Marincola and Lander 2014). The industry sponsored agreements are the second largest source of research funds in MRC, which are expected to bring a sum of USD 9874,000 to the center between 2015 and 2020 (Marincola and Lander 2014). SMRC collaborates with national (e.g., QBRI, QCRI, WCM-Q, QU, Supreme Council of Health and others) and international (e.g., Illumina Inc., The Jackson Laboratories, Stanford University and other) institutions. Up until 2015, the center published around 348 research papers (QF 2015). Nevertheless, as identified in the five year strategic plan of SMRC, major risks that the center face, similar to most other research institutes in Qatar, include lack of engagement of local enterprises, failure to achieve timely expected or impactful results and the operations of research institutions in silos (Marincola and Lander 2014). The center also offers scholarships and graduate associate programs to support and develop a culture of medical research within the country.

5.4.1.7 QMIC

QMIC, formerly known as Qatar University Wireless Innovation Centre (QUWIC), is a joint effort of QSTP and QU. Launched in 2009 as QUWIC and rebranded in 2012, QMIC operates independently and aims to use locally engineered innovations and knowledge to realize a market-focused innovation ecosystem in Qatar (Hariharan 2016). The center not only can assist Qatar in diversifying the economy through setting up technology-based industries but it can also help in promoting the culture of research and innovation among Qatar's youth with Internet of Things (IoT). QMIC offers a range of innovative solutions related to transport management, sensing network, fleet and asset management, environmental monitoring, and connected vehicles and drones. In addition, the center offers professional services to startups, entrepreneurs and SMEs in order to support their IoT initiatives. QMIC operates in partnership with the local stakeholders including MME, Ministry of Interior (MOI), QF, Vodafone, and Qatar Insurance Company. The R&D house of the center is committed to work with entrepreneurs, university students, and research centers at the early stage of the technology development (Hariharan 2016). Source of research funding of QMIC, like other research institutions in Qatar, is NPRP; the center was awarded 11 projects in the first 5 cycles of the National Priorities Research Program (NPRP)—a QNRF flagship funding program (QMIC 2012). In the fiscal year 2011–2012, QMIC published 3 journal papers, 11 conference papers, and 1 book, and filed a patent as well (QMIC 2012). More recent data on QMIC's publications and funding is not available.

5.4.1.8 QBB

QBB is a center operating under the flagship of QF since 2010. QBB is collaboration between HMC, Ministry of Public Health, and Imperial College London, which aims to enable national scientists to conduct medical research on prevalent health issues related to Qatar. QBB is responsible for the collection and storage of the biological samples (e.g., blood and urine) of Qatar's local population and the residents living in Qatar for more than 15 years. The data is subsequently used in biomedical research to develop medical treatment and disease prevention programs. The center makes this data available to other institutes for research under 'Research Access Projects'. According to an annual report of QBB, 35 research access projects were active in 2016 (QBB 2017). The institutions involved in the research access projects are SMRC, HMC, QBRI, QU and others. Furthermore, QBB collaborates with QNRF and Qatar Genome Project to support research and innovation in the area of personalized medical care through the Path towards Personalized Medicine grant. On the research impact side, in 2016, the researchers at QBB published 4 research papers and presented research in 5 conferences (QBB, 2017)—these numbers do not include the publications made under the research access projects.

5.4.1.9 QIC

QIC is a consortium of public and private organizations, established to engage the academic community with the entrepreneurs to enable innovation in the country. QIC, launched in March 2017, is a joint venture between the Ministry of Transport and Communications, Supreme Committee for Delivery & Legacy, QSTP and Ooredoo, along with 15 other national stakeholders.[1] The center launches various innovation programs and initiatives to support the home-grown innovation capabilities through collaborations, collective resource development, and stakeholder engagement. The primary focus of QIC, in the early phase after its conception, is to provide innovative solutions for the FIFA World Cup 2022 (to be hosted in Qatar); however, the strategic scope of QIC is broader than just an event. In order to accelerate the knowledge transfer through innovations and discoveries, QIC expects to launch innovation-driven entrepreneurship opportunities and joint public-private innovation projects in the future. The consortium is still at its early stage which is why details on governing mechanism, funding, and property rights are still not available.

---

[1] Al Jazeera, Aspire Academy, QBRI, International Chamber of Commerce, Josoor Institute, Qatar Airways, Qatar Business Incubation Centre, Qatar Chamber of Commerce, Qatar Development Bank, Qatar Financial Centre, Qatar Museums Authority, Qatar National Bank, Qatar Rail, Qatar Tourism Authority and QU.

5.4.2 Government Initiated Innovation Clusters

EC of QF, launched since 1997, is a knowledge cluster of global profile since it is a home to 8 international university branch campuses (6 American, 1 British, and 1 French) and 1 national university. QF has invested heavily in this project to import elite HEIs in Qatar with an aim of internationalization, massification and marketization of higher education in Qatar (Khodr 2011); the six U.S. branch campuses receive collectively more than $320 million each year (Barnawi 2017). In return, serious efforts are made by the members of the cluster to instigate IUGPs to promote research and innovation. Besides EC, the country is also undertaking a new initiative of development of special economic zones/clusters under Manateq, previously known as Economic Zones Company, owned by government of Qatar.

5.4.2.1 EC

Most entities under the flagship of QF are physically located in the cluster city, i.e., EC. Since the research institutions, innovation centers, and science park are in proximity, as shown in Fig. 5.2, EC serves as a hub of research and innovation in

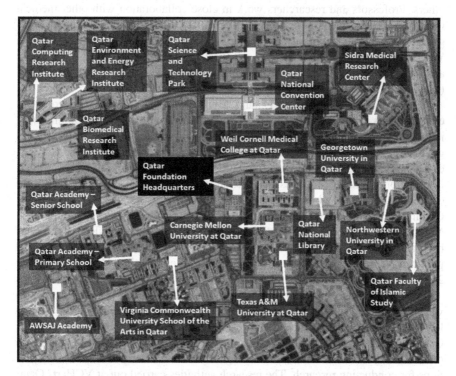

Fig. 5.2 EC landmarks (image taken from Google Maps—2017)

Qatar.

Texas A&M University at Qatar (TAMUQ) has been established in EC since 2003. The university offers Bachelor programs in chemical, electrical, mechanical, and petroleum engineering and a Master program in chemical engineering. TAMUQ is the spearhead of university-industry collaboration in EC and works together with its partners and customers which include Maersk Oil, ConocoPhillips, Chevron, QP, Schlumberger and others. The technical services department of the university provides services and expertise to local industries and organizations through consultation, training, and certification. In 2015, the university had research projects worth USD 196 million (The Texas A&M University System 2016). The outcomes of the research projects reported in 2016 include: 422 research papers, 9 books, 138 conference presentation, 11 inventions, and 75 invention disclosures (Texas A&M University at Qatar 2016).

Weill Cornell Medical College in Qatar (WCM-Q), established since 2001, offers six years medical degree programs leading to Doctor of Medicine. In addition to the world class instructional profile, WCM-Q is active in research relevant to the Qatari society. The university has 34 research laboratories specialized in various disciplines such as infectious diseases, cell signaling, genetics and immunology, and metabolomics and others. In addition, there are nine core facilities to support the research activities, including bioinformatics, genomics, imaging, proteomics and others. Professors and researchers work in close collaboration with other medical institutions, government ministries, and QSTP. As part of student development program, university offers internships, training and student research excellence awards. WCM-Q has published around 500 scientific papers since 2010 (up until 2017), and their publications were cited approximately 1900 times between 2010 and 2015 (Weill Cornell Medicine—Qatar 2017).

Carnegie Mellon University in Qatar (CMUQ), operating since 2004, offers Bachelor programs in business, computer science, information sciences, and bioinformatics. Even though CMUQ does not offer graduate program, the faculty involves undergraduate students in their research projects. The research program at CMUQ is interdisciplinary and revolves around computing and mathematics, biological sciences, information systems, economics, business and social systems. Research activities are carried out in close collaboration with the industry partners, which include Vodafone, Commercial Bank, KPMG and others, and government ministries, including MOI, MEC, and Ministry of Foreign Affairs. Since the inception of QNRF, CMUQ has been awarded USD40 million in grants. The university has published more than 50 publications in the academic year 2015–2016 (Carnegie Mellon University—Qatar 2016).

Virginia Commonwealth University School of the Arts in Qatar (VCUarts Qatar) was the first university to start operations in EC in 1998. The university offers Bachelors and Masters degree in fine arts and design. Although the core purpose of VCUarts Qatar is to offer instructional programs, the university also aims to collaborate with other education and research institutions, government ministries, and private institutions for conducting research. The research activities carried out at VCUarts Qatar are flexible, ranging from social science (e.g., migrant worker housing) to technical research (e.g., synthesis of nanomaterials capable of extending the lifespan of fragile documents).

Georgetown University in Qatar (GU-Q), established since 2005, offers a Bachelor program in foreign service with various specializations, and a certification program in political science. The university has recently announced an executive Master program in emergency and disaster management. Similar to VCUarts Qatar, GU-Q is an instructional university primarily, however, the faculty and students are involved in various research projects. Research activities carried out at GU-Q are in the field of international relations, political economy, and domestic politics of the GCC. The university has published more than 250 research papers, and about 45 books and 68 book chapters between 2005 and 2015 (Georgetown University—Qatar 2015).

Northwestern University in Qatar (NU-Q), established since 2008, offers Bachelors degree in communication, journalism and liberal arts. The university research division works in collaboration with its partners QCRI, Doha Film Institute, and Al Jazeera Media Network. The major source of funding for research is the NPRP. In the academic year 2015–2016, the university published 14 journal articles and 5 books (Northwestern University in Qatar 2016).

University College London Qatar (UCL-Qatar) is one of the youngest IBC in EC, operational since 2011. UCL-Qatar offers Masters degree in archaeology, conservation, cultural heritage and museum studies. The university research projects focus on developing a regional understanding of Arab and Islamic World. Since the university is relatively young, the impact of research outcomes is not significant.

HEC Paris is also operating in Qatar since 2011. The university offers a Masters degree in management, and advanced certification courses on Innovation and Social Business, Entrepreneurship and Innovation, Entrepreneurship-Project Accelerator, Leading Digital Transformation and others. While HEC Paris in Qatar is considered as an instructional institution, the university established its research office in mid-2014.

HBKU is a national university operating in EC since 2010. HBKU offers Bachelor, Masters and Ph.D. degrees in Science & Engineering, Health and Life Sciences, Islamic Studies, Humanities & Social Sciences, and Law and Public Policy. The intent of establishing a national university in cluster of IBCs was to synergize the universities' activities in EC and learn from the IBCs to enhance the national research capabilities (Hamad Bin Khalifa University 2016). One of the main objectives of HBKU is to support and increase the potential of commercialization of research outputs. HBKU has joint initiatives with various industry partners, including QP, Exxon, and Shell, and government ministries and agencies, including Ministry of Justice and MOI. In addition to departmental research, HBKU is mandated to monitor and boost R&D activities at the three national research institutes, i.e., QEERI, QBRI, and QCRI.

5.4.2.2 Manateq Special Economic Zones

Manateq is another initiative taken by the government of Qatar to diversify and transform its hydrocarbon driven economy to a knowledge-based economy (Manateq 2016). Manateq was first established in 2011 through the Minister of Business & Trade Decision No. 272. The aim of the economic zones was to develop and support a vibrant private sector in Qatar. Manateq Special Economic Zones is a project that undertakes the development of Special Economic Zones, i.e., strategic locations in three different parts of Qatar to encourage and facilitate the establishment of new domestic and international industries and businesses. One of the major attractions for investment in these zones is the tax exemption for approved businesses. The zones also offer duty free trade and provision of 100% overseas ownership of SMEs. In addition to the Special Economic Zones, development of Logistics & Warehousing Parks also comes under the scope of Manateq. The economic zones and logistics parks are expected to serve as the one-stop-shop for local and foreign businesses which are considering a setup in Qatar.

5.4.3 Science/Technology/Business Parks

QSTP, inaugurated in 2009, is more than USD 800 million investment of QF with an ambition of creating an innovation hub in the country and fostering the settings required for accelerating commercialization of research via entrepreneurship. The physical location of the only science park in Qatar, as shown in Fig. 5.2, provides a perfect environment for the IBCs and national research institutions to collaborate with each other and translate the research outcomes to commercial product and services. In order to support the research themes announced under QNRS, QSTP hosts various programs for technology development and startups, as shown in Fig. 5.3. The doors of QSTP are not only open for the national companies but institutions from around the world are encouraged to commercialize their technologies in Qatar.

QSTP's support programs are designed to provide entrepreneurs and startups with access to incubation, funding, training, and mentorship. The park offers office space, laboratories, access to technical and marketing facilities and free-trade zone, and commercialization support for its resident companies which include the incubatees (i.e., the startups in the incubation and residency program) and tenants (i.e., research centers of multinational corporations including ExxonMobil, Siemens, Total, Rolls-Royce, Vodafone and others). Between 2014 and 2015, QSTP was home to 38 resident companies (654 staff members working in R&D) and these companies invested a total of QAR 1.35 billion in research, development and innovation, which produced an output of 28 patents (QF 2015). The center serves as a core medium of engagement between the universities, industries and government in Qatar. QSTP collaborates with various entities and stakeholders in research projects of national importance. Some of these collaborators include Chevron, Shell, QP, General Electric, Qatar

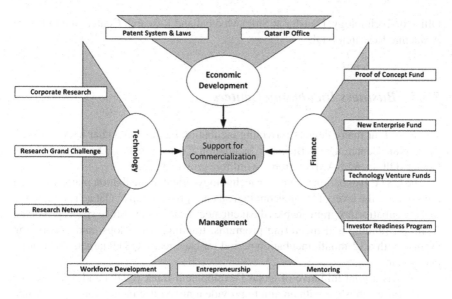

Fig. 5.3 Framework of QSTP (Roberts 2010)

Airways, Air Bus, Imperial College London, TAMUQ, The University of Sheffield, QU, HMC, MOTC, MOI, US DOE, and European Innovation Academy.

5.4.4 Technology Transfer Offices

Office of Research Strategy and Impact Management (RSIM) administers intellectual property and technology transfer activities in Qatar (QF R&D 2017). Formerly known as Intellectual Property and Technology Transfer (IPTT) office, RSIM was established in 2014 with an aim to oversee the management, acquisition and monetization of IP. RSIM has the mandate to capture the advances, discoveries and inventions associated with QF. The centralized office transforms these advancements into commercial opportunities which can enhance Qatar's global reputation and attract foreign investments. Not only for the QF, RSIM also manages and commercializes the original technologies of QP-Research, QU and HMC.

RSIM offers a special program, Al-Khabeer, to assist local startups and inventors in protecting their ideas and building credible business plans and models. Currently, the office manages more than 400 inventions at different market readiness levels (QF R&D 2017). RSIM develops strategies and research targets for Qatar's research institutes through identification of technology hotspot (to target specific research areas). The office seeks to build a network of potential licensees and investors for commercialization and transfer of technology. Other IP and technology transfer related offices in Qatar are Intellectual Property Rights Protection Department under MEC,

Office of Technology Transfer at Sidra Medical and Research Center, and Office of Academic Research—QU.

5.4.5   Business Incubation Centers

There are four major centers providing incubation facilities in Qatar namely QSTP Incubation Center, Qatar Business Incubation Center (QBIC), Digital Incubation Center (DIC), and NAMA (Arabic word for 'growth') center.

The QSTP Incubation Center is a technology-oriented incubation platform which aims to raise the level of local technological entrepreneurship in Qatar. The selection to the incubation program enables the incubatees to access the office space, prototype facilities, funding and mentoring programs, training, workshops, and networking events. With a 12-month incubation period for each team, QSTP targets 20 projects every year.

QBIC is a joint initiative of Qatar Development Bank (QDB) and NAMA. The center was established with an aim to provide a mixed-use incubation center which can boost the entrepreneurs and companies in the startup as well as can propel the existing SMEs to grow their businesses (QBIC 2017). Established since 2014, QBIC has a mission to *"develop the next QAR100 million companies in Qatar"*. The center runs two incubation programs per year, i.e., LeanStartup and LeanScaleup, and is currently a home to 59 incubated companies. In addition to office space, workshops, and coaching and mentoring services, QBIC also offer financial assistance. Seed funding of up to QAR100,000 is available for validation and prototype development whereas an equity finance of up to QAR300,000 is available for growth of business. Loan opportunity for scale-ups can go up to a value of QAR4 million through QDB. QBIC has made a cumulative investment of QAR3 million in the startups up until 2017 and the revenue generated by those startups reached QAR21.4 million, with 96% incubation occupancy rate. QBIC operates in partnership with Qatar Tourism Authority and Ooredoo to focus especially on the startups and SMEs that intend to diversify Qatar's economy in tourism and digital solutions' sector. Knowledge partners of the center include Qatari Business-Women Association (QBWA), QMIC, Microsoft, and Qatar Finance and Business Academy (QFBA).

Inauguration of DIC is a milestone for innovation culture in Qatar since it is the first center to be launched directly by a government ministry, i.e., MOTC. DIC was founded in 2011 with a mission to advance ICT innovation in Qatar. DIC's sectors of interest for technological innovation include education, government services, health, transportation and logistics, urban planning, environment, financial services, energy, construction and others. DIC works in close partnership with Vodafone, KPMG, BusinessPulse, and Amazon Web Services to ensure that technological innovations can be commercialized to support businesses in Qatar. The center aims to assist young entrepreneurs in transforming innovative ideas into viable businesses through incubation programs. DIC offers two tracks for incubation: (a) the Launchpad— 6 months track of developing prototype from an original tech idea; and (b) Start-up track—2 years program to support the young tech-entrepreneurs and start-ups in developing market-ready products. Even though the center does not offer direct financial support the incubatees may receive indirect support of up to QAR400,000 from QDB and other financial institutions depending upon the market readiness level of the product or service. The successful start-ups which have potential to grow are expected to be hosted as tenants in Manateq Special Economic Zones. The center has incubated a total of 29 companies since its launch in 2011.

NAMA, formerly known as Social Development Center, was established in 1996 under the patronage of Sheikha Mozah Bint Nasser, wife of the then Emir, with an aim of social development in Qatar (NAMA 2017). Later in 2013, the center became a member of the QF for Social Affairs. In addition to community engagement and services, the center has an entrepreneurship wing which provides business incubation, financial support, and training and consultation services for the development of social entrepreneurs. The incubation facility at NAMA helps the existing and potential entrepreneurs of small and micro businesses in assessing the feasibility of concepts, developing business plans, and establishing communication channels with clients. The incubatees also receive legal advice and financial support, depending on the market readiness level of their services. NAMA operates in collaboration with its partners, which include QDB, ExxonMobil, Qatar Olympic Committee and others. Other than the entrepreneurship wing, the center offers school and university education support, and professional development programs.

5.5 Support Programs

Research organizations, industries and government bodies in Qatar partner in number of programs to facilitate idea creation and translation. Some of the support programs in Qatar have been listed in the following subsections.

5.5.1 Public Procurement

In order to align Qatar's public procurement law with international best practices, government of Qatar superseded Law No. 24 of 2015 (Regulation of Tenders and Auctions) over Law No. 26 of 2005 (Promulgating the Tenders and Bids Regulatory Law). The most significant change as a result of this amendment is the decentralization of public procurement. The other major change which came in effect with the suspension of previous law is the waiver (partial or full) of performance bonds and guarantees for public procurement from the SMEs. The new law guarantees 30% domestic procurement in the contracts of national projects. The SMEs can provide the financial assurance more conveniently through QDB. This relaxation will encourage the SMEs to increase their participation in public bidding and tendering process, and thus, will provide innovative means for the diversification of national economy.

5.5.2   Tax Incentives

As per the Law No. 36 of 2005, QSTP is a free-trade zone which makes the park an attractive location to establish foreign-owned technology-based business in Qatar. Some of the benefits of working in a free-trade zone include 100% ownership of the company, trade without the involvement of local agents, no taxes, duty free relevant imports, and unrestricted repatriation of profits (QSTP 2015). In addition to the tax-free zone in QSTP, tax exemption at the Manateq Special Economic Zones (for the approved businesses) is another example of attracting foreign investment through tax incentives and strategically aligning that investment with Qatar's new economic perspective.

5.5.3   Internship, Training, Entrepreneurship and Innovation in Curriculum

Conventional internship opportunities in the petrochemical industry are frequently available to the undergraduate and graduate level students in Qatar. Nevertheless, with an increasing emphasis on innovation and entrepreneurship in the country, students get a chance to be exposed to innovation centers during their study. This includes internship opportunities with QCRI, QBRI, QMIC, and Siemen's Intern Innovation Center in Qatar among others. Medical Education Department of HMC also offers a wide range of graduate level medical programs such as the internship, residency, educational, training and scholarships with an aim to ensure its medical workforce to receive latest knowledge and state of the art trainings. Universities in Qatar have also played their part and included innovation and entrepreneurship studies in the educational curriculum at the graduate and undergraduate level. Notably, these courses are offered at CMU-Q, HBKU, QU (Center for Entrepreneurship), and HEC Paris-Qatar.

5.5.4   Public-Private Partnership Programs

R&D programs in Qatar have placed increasing importance on public-private partnerships more recently. For example, QNRF has revised its existing R&D programs and has introduced new programs to facilitate the public-private and industry-university-government partnerships. Similarly, HMC's Academic Health System (AHS) program is supporting knowledge sharing between public and private entities in the health sector. Also, there are financial and non-financial partnership programs offered by government and non-government entities to support entrepreneurs, start-ups and existing SMEs. For example, QDB has established various specialized programs to promote SMEs and entrepreneurs in Qatar including Al-Fikra-Qatar National Business Competition, JAHIZ-1 and JAHIZ-2, SME Equity Program, and TASDEER.

Several non-profit organizations such as Injaz Qatar, Silatech, Bedaya Center for Entrepreneurship and Career Development, and Roudha Center for Entrepreneurship and Innovation are providing specialized support to promote domestic businesses. QSTP and QDB also offers other advisory and funding support to startups and accelerators. Beside these specialized programs, the government HMC organizes interactive events and programs to facilitate networking and development of knowledge-based culture within the country.

5.5.4.1 QNRF

QNRF, established in 2006, operates as a member of QF community. QNRF's mission is *"to advance knowledge and education by providing funding opportunities for original, competitively selected research and development at all levels and across all disciplines"* (QNRF 2017). QNRF supports and funds research projects designed to address the grand challenges of Qatar, which include: Energy and Environment; Computer Sciences and ICT; Health and life Sciences; and Social Sciences, Arts and Humanities. QNRF offers various funding opportunities; the list of some of these programs have been tabulated in Table 5.2 to highlight the number of projects and amount of funding being awarded by QNRF from 2006 to 2015. Among these programs, the most notable is the NPRP which has funded more than 850 research projects worth US$737 million during the first seven years. This generous funding provided an opportunity to the universities operating in Qatar to collaborate with 445 international institutions from 49 different countries in various research projects. Starting 2017, the program has made industrial participation mandatory in order to promote industry-based research in Qatar (QNRF 2017b).

In addition to these focused programs, QNRF offers numerous other special programs, funding opportunities, and competitive awards to complement its vision of fostering a research culture in Qatar. A list of these special programs, with their respective objectives, is tabulated in Table 5.3.

More recently, QNRF has collaborated with Qatar National Library (QNL) to establish a research outcome repository, Research Outcome Center Search (ROCS), to advance the dissemination of knowledge in Qatar. ROCS will provide access to the results of QNRF-funded research projects to the members of QF community, QU, and all other networks that can access QNL database.

5.5.4.2 AHS

In 2011, HMC transformed the status of its hospitals to AHS in an effort to promote research and innovation in healthcare sector in Qatar (HMC 2017). HMC works in close collaboration with WCM-Q, University of Calgary-Qatar, QBRI, QU, and SMRC. The aim of the initiative is to integrate health, education, and research through collaborations between each partner institution in pursuit of new discoveries. The expected positive outcomes in the long run include enhanced patient care, advance-

Table 5.2 QNRF funding programs between 2006 and 2015 (Boukhris 2015; QNRF 2017b)

Program name	Program objective	Starting year	Number of cycles	Awarded projects	Funding (USD million)
Research programs					
National Priorities Research Program (NPRP)	To select research projects that address national priorities through supporting basic and applied research as well as translational research/experimental development	2007	7	866	737
National Priorities Research Program—Exceptional Proposals (NPRP-EP)	To provide extra funding opportunity for proposals that require additional investment in excess of the normal NPRP funding	2011	7	8	36
Capacity building and development programs					
Undergraduate Research Experience Program (UREP)	To stimulate a broad array of undergraduate research opportunities within Education City and Qatar University through faculty- and other researcher-led projects involving one or more students.	2006	16	761	23.8
Junior Scientist Research Experience Program (JSREP)	To foster a research culture in Qatar through supporting junior scientists to conduct research that is aligned with QNRS	2010	6	27	6
Secondary School Research Experience Program (SSREP)	To nurture research culture among Qatar's youth	2010	4	184	Not available

(continued)

Table 5.2 (continued)

Program name	Program objective	Starting year	Number of cycles	Awarded projects	Funding (USD million)
Graduate Sponsorship Research Award (GSRA)	To support human capital development by funding outstanding prospective students to undertake graduate research-based studies in a domestic or international approved university	2013	2	15	7
Postdoctoral Research Award (PDRA)	To encourage the recent Ph.D. graduates and young researchers at an early stage of their research career	2015	1	13	3.7
Grand total				1874	813.5

ment in medical research, community engagement, and economic and health benefits to the community. In order to stimulate innovation, AHS has established Innovation Awards. The aim of this initiative is to capture the fruitful ideas which do not get to the stage of proof-of-concept due to lack of funding and support and poor idea management. In addition, researchers at AHS can benefit from the QNRF funding as well.

5.5.4.3 QDB

QDB, previously known as Qatar Industrial Development Bank, was founded in 1997 with an aim to: (1) grow the country's private sector; (2) to diversify its economy; and (3) transform into a sustainable knowledge-based economy. In November 2014, Enterprise Qatar (EQ), a company established in 2008 to develop and promote SMEs and entrepreneurs in Qatar, was merged in QDB to centralize the support system for development of private sector. Since then, the bank has established various financial and non-financial programs to support entrepreneurs, startups, and SMEs in the various growth stages.

- Al-Fikra (Qatar national business competition)—Launched in 2013, Alfikra is a joint initiative of EQ, MIT Enterprise Forum, Silatech, and CMUQ. The primary objective of the competition is to educate and coach young business entrepreneurs (Al-Fikra 2017). The secondary objective is to encourage the establishment of new companies and foster a culture of entrepreneurship in Qatar. The competition

Table 5.3 List of special QNRF programs (QNRF 2017b)

Name of special programs	Program objective	Starting year	Support and funding
Conference and Workshop Sponsorship Program (CWSP)	To connect researchers in Qatar to their peers and colleagues, both at home and abroad, by exposing researchers or students to new research directions, findings, and education techniques	2009	Evaluated on case basis—usually USD 50,000 for each such event
Qatar-UK Research Networking Program (Q-UKRNP)	This joint research networking program is designed to provide financial support to bring together a Qatar-UK bilateral cohort of early career researchers to take part in workshops focusing on building links for future collaboration and enhancing the researchers' career opportunities	2013	Up to USD 50,000 for conducting workshop
Qatar Innovation Promotion Award (QIPA)	To support innovative ideas with high potential benefits to Qatar's economy and society	2014	Financial support of up to USD 100,000 (over 12 months)
Challenge 22—Innovation Award	Annual award established in collaboration with Supreme Committee for Delivery & Legacy to promote a culture of innovation in the Middle East and attract entrepreneurs, scientists and pioneers from across the Arab world to contribute towards the hosting and organization of major events like the 2022 FIFA World Cup Qatar	2015	USD 15,000 and an opportunity to win up to USD 100,000 after incubation. Also, networking and showcasing opportunity on an international platform

(continued)

Table 5.3 (continued)

Name of special programs	Program objective	Starting year	Support and funding
Belmont Forum	QNRF has joined the Food-Water-Energy Nexus consortium of Belmont Forum—the partnership of 27 major international funding agencies. With this collaboration QNRF seeks to foster research and innovation at the intersection of food, water, energy sustainable consumption and production, and urbanization	2015	Total funding for collaborative projects is approximately USD 1.5 million
Path Towards Personalized Medicine Presentation (PPM)	To support and advance research that aims to provide medical treatment tailored to the individual characteristics of each patient, based on her or his genetic profile	2015	up to USD 200,000 per year (up to 3 years)
Best Representative Image of an Outcome (BRIO)	To encourage researchers to tap into their artistic side by using stimulating images and visuals to communicate the outcomes of their QNRF-funded research to a wider audience, including the general public.	2015	not available
National Science Research Competition (NSRC)	This program, established in collaboration with The Ministry of Education and Higher Education, aims to encourage all middle and secondary school students in Qatar to utilize their learning to initiate contributing to knowledge production, creativity and problem solving	not available	A chance of nomination to the Education Excellence Day Awards and a chance to represent Qatar on an international platform

provides support to start-ups and entrepreneurs to develop business ideas that can be turned into successful private businesses. The winning idea of the competition receives financial award, incubation services, and vouchers for subsidized professional services. The net worth of the prize is about QAR8,000,000 (Al-Fikra 2017).

- QDB has taken the initiative to boost SMEs by providing ready-built industrial facilities (for rent) for manufacturing in different sectors at a competitive rate of 5QAR per square meter per month (QDB 2017a). The JAHIZ-1 program, launched since 2015, supports manufacturing of paints, metal coating, paper and paper board, domestic appliances, cutlery and general hardware, whereas JAHIZ-2 program, launched since 2016, is dedicated for food and beverage business. QDB also assists the manufacturers through a comprehensive start-up business solution which includes consultations, business plan, and feasibility studies. Projects related to innovative and environment-friendly products are given priority in the selection process.

- The SME equity program of QDB aims to provide capital financing to foster entrepreneurship and create sustainable returns (QDB 2017b). The program invests in innovative small enterprises and high-growth potential medium enterprises, led by Qatari nationals, in order to support domestic value creation in the country. The program can support small and medium enterprises with QAR1.5–7 million and QAR 3.5–18 million over a period of 8–10 years, respectively. The overall size of the program is QAR365 million. However, the program is highly competitive and proposals usually have to go through two tier assessment process before qualification.

- Launched in 2011, TASDEER (Arabic of 'export') seeks to support Qatar's domestic products in the international market through export development and promotion (QDB 2017b). TASDEER offers various workshops, training modules, and seminars to assist Qatar's manufacturing companies in becoming export ready. The program also helps domestic SMEs in identifying the in-demand products and their respective target markets based on international market studies. The domestic products are promoted at international exhibitions by organizing pavilions and through matchmaking between Qatar's SMEs and international buyers.

QDB offers various other services and programs in its effort to help Qatar in diversifying its economy. Some of these include:

- Ithmar—equity capital to Qatar's local entrepreneurs (financial support up to 90% of the project)
- Al-Dhameen—letter of guarantee to other financing banks (promising to repay up to 85% of the allocated funds, not exceeding QAR15 million in order to support bidding)
- Direct Lending—flat loans at competitive rates
- Jadwa—feasibility assessment of business ideas (50% of cost covered)
- Oqood—professional legal consultancy (50% of cost covered)
- Tadqeeq—accounting and auditing services (50% of cost covered)

- Marketing and Public Relations—planning and implementing marketing and public relations activities (50% of cost covered)
- Eyada—restructuring advisory to optimize sales, efficiencies and costs, and maximize profitability (50% of cost covered)
- Capability Development—various training and mentoring programs.

5.5.4.4 Injaz Qatar

Injaz Qatar, established since 2007, is a non-profit organization that works together with the local business community, corporate volunteers and educators to offer educational programs about workforce readiness, entrepreneurship, and financial literacy. These programs aim to develop business leadership and entrepreneurial skills in the youth at an early age (12–24 years) and therefore, targets an audience ranging from elementary school to university graduates. An example of such programs is the twelve session long 'Company Program', stretched over 4 months, for students from 14 to 22 years of age (Injaz Qatar 2017). The program enables the students to go through the complete life cycle of startup venture, including generation of business idea, team formation, raising capital funding, devising business plans, and producing and selling the product or service. The program concludes with a national and regional level competition, called Mubadara, for the best company award. Since Injaz Qatar is a non-profit organization, it receives support and funding from its partners and collaborators which include NAMA, Qatar Stock Exchange, ExxonMobil, Boeing, Bechtel, Standard Charted Bank, Vodafone, Siemens, QU, MOE and others.

5.5.4.5 Silatech

Silatech, launched since 2008, is an initiative of Sheikha Mozah Bint Nasser, the wife of the then Emir, and is fully supported by the state of Qatar. The organization was established with an aim to create jobs and economic development opportunities for Arab youth throughout the world. As of March 2017, Silatech raised USD237 million through a network of 150 co-partners and connected more than 434,000 young Arabs to job opportunities in 16 Arab countries (Silatech 2017). The organization works in close collaboration with governmental and non-governmental institutions, educational and financial institutions, and private sector. Silatech equally supports the enterprise development initiatives taken by young Arabs. In addition to financial backing, Silatech offers non-financial support through capacity building and technical assistance programs. Silatech encourages other financial institutions to design loan products to support Arab entrepreneurs. The major difference between Silatech and other organizations, of similar stature in Qatar, is that the former engages in applied research to make its programs more effective, and also to assess government policies relating to job creation and economic development of Arab youth. Through policy program, Silatech engages with the policy makers directly in different countries to achieve wider impact.

5.5.4.6 Bedaya Center for Entrepreneurship and Career Development Programs

Bedaya Center for Entrepreneurship and Career Development, established since 2011, is a joint effort of QDB and Silatech with an aim to provide Qatar's young generation access to wide range of services, including career counseling, employability skills development, entrepreneurship, and networking events. The center has established social engagement platforms which promote career awareness and development among Qatar's citizens, including men and women. On the entrepreneurship side, Bedaya center has separate programs for fostering entrepreneurial culture among the students (boot-camps, workshops, and business enterprise challenge) and enhancing the business productivity of early stage startups (through mentoring, networking, and setting up startup markets). The center is recognized in the region for hosting competitive entrepreneurial events such as 'Get in the Ring', where new startups are provided a platform to pitch ideas in front of local and regional supporters. In 2016, the center reached a total of about 13,000 students in 9 universities and 38 schools. It supported 251 startups from idea to implementation stage, and carried out 102 training programs (Bedaya Center 2016). The supporting partners of Bedaya Center are MOE, QF, QU, CMUQ, Shell Qatar, Qatar National Bank, DIC, QBIC and others.

5.5.4.7 Roudha Center for Entrepreneurship and Innovation

Roudha Center, founded in 2011, is non-governmental and not-for-profit organization which operates as a one-stop shop for youth and Qatari women entrepreneurs with innovative business ideas (Roudha Center 2017). In addition to the support services, such as training programs, the center also offers incubational assistance to boost the number of women entrepreneurs in Qatar. The sponsors and partners of the center include Silatech, Bedaya Center, Injaz Qatar, VCU-Q, and ExxonMobil among others. As of 2014, Roudha center helped about 6000 women in starting their own businesses in Qatar (Changemakers 2017).

5.5.4.8 Other Support Programs for Start-Ups and Accelerators

QSTP provides various funding opportunities to meet the expectations of tech-based firms at different levels (i.e., from seed funds to growth-stage):

- New Enterprise Fund—A total of USD30 million is available to provide support for the start-up of technology companies in Qatar (USD500,000 to USD3 million each startup).
- Tech Venture Fund—In order to support the tech-based companies in Qatar, QSTP offers funding ranging from USD 50,000 to USD3 million.
- Proof of Concept Fund—USD100,000 to USD500,000 is offered to support the innovations that are technically and commercially viable.
- Product Development Fund—This QAR1.4 million fund supports development of products and services which are relevant to the local market needs.
- Technology Development Fund—QNRF manages this fund to provide gap funding for new projects at QSTP. The objective of technology development fund is to increase the commercialization potential of the new projects. The amount of funding is determined through the readiness level of the technology.
- Corporate Research Fund—QF offers this technology development fund to enhance corporate research and commercialization at QSTP.

Beside funding, QSTP runs various other specialized support and accelerator programs to fulfill its mission of developing and enhancing a culture and ecosystem for innovation, research and entrepreneurship.

- Startup in Residence Program—This is a special program offered by QSTP to attract international technology startups who wish to establish their presence in Qatar and who can provide solutions to the challenges of the country. QSTP select only those startups in this program which have a strategic rationale for presence in Qatar.
- Arab Innovation Academy (AIA)—The two weeklong boot-camp which is a joint initiative of QSTP and European Innovation Academy (EIA) provides a vibrant startup experience to its participants—from idea to the tech venture startup.
- QSTP XLR8—This accelerator program helps in the 'idea stage' to determine the commercialization potential. The three-month long program offers training, coaching, and mentorship to improve the innovative ideas. Three coaching modules are designed to support teams from ideation to minimum viable product, including Product-Market Fit, Customer Traction, and Investor Awareness.
- Doha Dojo—QSTP runs this growth accelerator program, in partnership with '500 startups', to prepare limited number of startups in scaling-up their operations. Growth mentors help the 'product market fit' startups in areas such as customer acquisition, growth hacking, and pitching to investors.
- Innovation Mindset (Education)—QSTP offers the Graduate level students, studying in the local universities, a chance to visit technology intensive centers, such as Silicon Valley, and EIA. QSTP targets its mission of fostering innovation culture among the young local student thorough the Student Innovation Trips.
- Research-To-Startup (QF R&D)—This is a QF sponsored program run by QSTP in collaboration with Wasabi Global Ventures, QCRI, and RSIM. It provides a smooth launch-pad for tech-entrepreneurs who aim to commercialize the technologies developed at research institutions and universities in Qatar.

5.5.4.9 Interactive Programs

The government of Qatar also hosts several interactive programs at national and international level to promote the culture of research and innovation on one hand and to facilitate networking and collaborations on the other. Following are some selected examples of such programs:

- QITCOM Exhibition and Conference—QITCOM is Qatar's biggest digital event organized by MOTC. The event serves as a platform to connect international experts, government officials, industry specialist, business associates, academics, innovators, entrepreneurs, and investors on annual basis since 2011. The event hosts various activities and competitions including coding and programing competition, startup-investor matchmaking, workshops to promote innovation, IT business awards, and digital youth festival. The 2017 edition of QITCOM involved 70 public and private partners, 15 sponsors, around 120 local and global exhibitors, over 20 visiting international delegations, more than 13,000 visitors, about 300 entrepreneurs and more than 50 local, regional and international startups (MOTC 2017).
- Qatar Science, Technology, Engineering, Art and Math (STEAM) Fair—Qatar STEAM Fair is an annual science, technology, engineering, arts, and math school competition organized by QF R&D in collaboration with QNRF and the Supreme Education Council. This nation-wide competition allows school students to present their projects in front of industry experts for a cash prize of USD20,000. The winners also get a chance to represent Qatar at a renowned international science competition, i.e., Intel International Science and Engineering Fair (QF 2016b).
- Stars of science—Started in 2009 by QF, Stars of Science is a TV show which allows aspiring science and technology entrepreneurs from the Arab world to develop creative solutions for regional problems in a 'reality show' format. A technical team of experts choose the candidates for the show based on their online applications. The selected candidates present their innovations to a panel of expert judges who assess and eliminate candidates based on prototyping and validation of ideas. The four finalists compete with each other for a share of USD600,000 in seed funding which is awarded by the jury and through online voting (QF 2017). The ownership of the innovation remains with the participating candidates.

A summary of the chapter has been tabulated in Table 5.4.

Table 5.4 Summary of the IUGP trends and drivers in Qatar

	1960 and before	1960–1970	1970–1980	1980–1990	1990–2000	2000–2010	2010 and after
Governance:	MOE		HMC, MME		QF, QDB	MEC	MOTC
Legislations:						Trademark Law, Law for Protection of Trade Secrets, Law for Establishment of Free Zone for QSTP, Patent Law	Law for Establishment of Qatar Development of SMEs
Intermediaries:							
Research Institutes/centers/consortiums					MRC	QMIC, Injaz Qatar, Silatech	QCRI, QEERI, QBRI, iTRI, SMRC, QBB, QIC, Bedaya Center for Entrepreneurship and Career Development, Roudha Center for entrepreneurship
Clusters					EC		Manateq Special Economic Zones
Science Parks/Business Incubators/TTOs					NAMA Center	QSTP, QSTP Incubation Center	QBIC, DIC, RSIM
Policies:							

(continued)

Table 5.4 (continued)

	1960 and before	1960–1970	1970–1980	1980–1990	1990–2000	2000–2010	2010 and after
Public procurement							Revised Tenders and Bids regulations
Tax incentive						QSTP Tax Free Zone	Tax free Manateq Special Economic Zones
Academic entrepreneurship and innovation programs							Academic innovation and entrepreneurship Program
Partnership programs						NPRP, UREP, CWSP, QSTP funding, start-up and accelerator programs, Interactive Programs	NPRP-EP, JSREP, SSREP, GSRA, PDRA, Q-UKRNP, QIPA, Challenge 22-Innovation Award, PPM, Belmont Forum, BRIO, AHS, Alfikra-Qatar National Business competition, JAHIZ program, SME Equity Program, TASDEER

References

Abduljawad, H. (2013). *University–Industry–Government partnership: A state of Qatar transformation to a knowledge-based economy case study.* Bellevue University.

Abduljawad, H. (2015). Challenges in Cultivating Knowledge in University-Industry-Government Partnerships—Qatar as a Case Study. *The Muslim World.* Retrieved from http://onlinelibrary. wiley.com/doi/10.1111/muwo.12080/full.

Al-Fikra. (2017). Al-Fikra, The national business competition. Available at: http://www.alfikra.me/ en/. Accessed on January 5 2018.

Barnawi, O. Z. (2017). The architecture of a neoliberal english education policy in Qatar. In *Neoliberalism and English Language Education Policies in the Arabian Gulf.* Routledge.

Bassil, A., Stolworthy, J. C., & Boukhris, O. (2017). *Diversifying Qatar's Economy through Science &Technology Investments—Successes, Challenges & Opportunities,* p. 17.

Bedaya Center. 2016. Bedaya Center achievements in 2016. *Bedaya.* Available at: http://www. bedaya.qa/en/#we-helped-a-lot. Accessed online on October 10 2017.

Boukhris, O. (2015). *Qatar National Research Fund,* p. 23.

Carnegie Mellon University—Qatar (2016). *Annual Report 2015–16,* Doha.

Changemakers. (2017). Women powering work: Innovations for economic equality in the mena region. Available at https://www.changemakers.com/MENAwomen/competition/entries/roudha-center. Accessed on January 5 2018.

Cornell University, INSEAD, WIPO. (2017). *The global innovation index 2017—Innovation feeding the world.* Ithaca, Fontainebleau, and Geneva.

Elsheshtawy, Y. (2008). The evolving Arab city. *Tradition, Modernity and Urban Development.* Retrieved from http://www.academia.edu/download/30695941/The_evolving_Arab_city_ch-10. pdf.

Georgetown University—Qatar. (2015). *Georgetown University School of Foreign Service in Qatar—Research Report 2005–2015,* Doha.

Hamad Bin Khalifa University. (2016). *Hamad Bin Khalifa University Strategic Plan 2016—2026,* p.40.

Hariharan, S. (2016). Qatar Ushers In The IoT Era: R&D Center QMIC Leads Region's Smart Mobility Research. *Entrepreneur.* Available at https://www.entrepreneur.com/article/276282.

HMC. (2017). Academic health system. Available at https://www.hamad.qa/EN/Education-and-research/Academic-Health/Pages/default.aspx. Accessed on January 5 2018.

INJAZ Qatar. (2017). Company program. Available at: https://www.injaz-qatar.org/. Accessed on January 5 2018.

Khodr, H. (2011). The dynamics of international education in Qatar: exploring the policy drivers behind the development of education city. *Journal of Emerging Trends in Educational Research and Policy Studies, 2*(6), 514–525.

Manateq. (2016). *Prime foundations—boundless opportunies,* p. 23.

Marincola, F. M. & Lander, H. (2014). *Five year strategic plan for research at sidra medical and research center,* p. 61.

MDPS Qatar. (2016). *Ministry of development planning and statistics.* Available at http://www. mdps.gov.qa/en/statistics1/pages/topicslisting.aspx?parent=Population&child=Population.

MDPS Qatar. (2017). *Finding of research & development survey in the state of Qatar 2015.* Ministry of Development Planning and Statistics, April 2017.

Merekhi, N. (2013). Research forward. In: *Qatar national statistics day forum* (p. 16). Doha: Qatar Foundation.

Ministry of Education and Higher Education. (2017). List of Universities in Qatar. Retrieved June 8, 2017, from http://www.edu.gov.qa/En/ServicesCenter/Pages/Supported-Universities.aspx.

MOTC. (2017). QITCOM 2017 Opening. Available at http://www.motc.gov.qa/en/news-events/ news/hh-emir-be-patron-qitcom-2017-opening-March-6. Accessed on January 5, 2018.

NAMA. (2017). http://www.sdc.org.qa/about. Accessed on June 16 2017.

Northwestern University in Qatar. (2016). *2016 Year in review—Achieving academic aspirations*, Doha.

OPEC. (2016). Qatar facts and figures. Available at http://www.opec.org/opec_web/en/about_us/168.htm. Accessed on June 8 2017.

QBB. (2017). *Qatar biobank annual report 2016/17*, Doha. Available at http://www.qatarbiobank.org.qa/app/media/1730.

QBIC. (2017). http://www.qbic.qa/about/. Accessed on June 16, 2017.

QBRI. (2017a). http://www.qbri.org.qa/about-us/overview. Accessed on June 15 2017.

QBRI. (2017b). Current research collaborations. Available at http://www.qbri.org.qa/partnerships/current-collaborations. Accessed on June 15, 2017.

QCRI. (2014). Join the Innovation., p. 6. Available at http://www.qcri.qa/app/media/3827 (Accessed January 5, 2018).

QCRI. (2017a). http://qcri.org.qa/about-qcri/vision-mission-and-values. Accessed on June 15, 2017.

QCRI. (2017b). QCRI—partners and collaborators. Available at http://qcri.org.qa/about-qcri/partners-and-collaborators. Accessed on 15 June 2017.

QDB. (2017a). Engineering projects. Available at: https://www.qdb.qa/en/engineering-projects. Accessed on January 5, 2018.

QDB. (2017b). Products and services. Available at https://www.qdb.qa/en/products-services. Accessed on January 5, 2018.

QEERI. (2017a). http://www.qeeri.org.qa/en/about-us/overview. Accessed on June 15, 2017.

QEERI. (2017b). QEERI—local and international partners. Available at http://www.qeeri.org.qa/en/about-us/partners. Accessed on June 15, 2017.

QF. (2015). *Annual Report 2014–15*, Doha.

QF. (2016a). *Annual Report 2014–15*, Doha.

QF. (2016b). *Young scientific minds impress at Qatar STEAM Fair*, p. 4. Available at https://www.qf.org.qa/app/media/44822. Accessed on January 5 2018.

QF. (2017). About the show. Available at: http://www.starsofscience.com/about-show. Accessed on January 5, 2018.

QF R&D. (2017). *Intellectual property: Qatar foundation*. Available at https://www.qfrd.org/en-us/Intellectual-Property. Accessed October 10, 2017.

QMIC. (2012). *QMIC Annual Report 2011–12*.

QNRF (2017a). *Qatar Nation Research Fund—Vision and Mission*.

QNRF. (2017b). Funding. Available at: https://www.qnrf.org/en-us/. Accessed on October 10, 2017.

QSTP. (2015). *Free zone regulations*, p. 69.

Roberts, E. (2010). QSTP: Fostering research and commercialisation to grow Qatar's knowledge economy, p. 19.

Roudha Center. (2017). Roudha center—Our story. Available at http://www.roudha.org/main_data_more.php?main=4. Accessed on January 5, 2018.

Silatech. (2017). *Annual report 2016–2017*. Available at http://www.silatech.org/docs/default-source/publications-documents/annual-report-2016-2017.pdf?sfvrsn=10.

Texas A&M University at Qatar. (2016). *Research performance 2016*. Available at https://www.qatar.tamu.edu/research/research-capabilities. Accessed on October 10, 2017.

The Texas A&M University System. (2016). *Fact Sheet 2015–16*, p. 17. Available at http://assets.system.tamus.edu/files/communications/pdf/Facts2015.pdf.

The World Bank. (2017). https://data.worldbank.org/. Accessed on December 17, 2017.

Weill Cornell Medicine—Qatar (2017). *Fact Sheet 2016–17*, p. 2.

Chapter 6
Comparison Between the IUGP Settings and Global Innovation Index of Qatar, United States, Norway, and Singapore

Abstract In the first part of this chapter a qualitative comparison of the IUGP enablers is performed between US, Norway, Singapore, and Qatar, which were discussed in detail in Chaps. 2–5. The qualitative comparison reveals that the IUGP history in the US, as one might expect, is the longest in comparison to the other countries. At the same time, Qatar is still at an engagement level in terms of its IUGPs and overall innovation system. In the second part of this chapter, to further explore the specific areas where Qatar has a room to improve (in comparison to the other three countries), a comprehensive quantitative comparison of Global Innovation Index (GII) and its indicators is performed between the four countries (Cornell University, INSEAD, WIPO, 2017). The comparison highlights that Qatar can enhance its innovation system and IUGP settings through improvements in regulatory environment, research and development, market sophistication (such as credit and investment), knowledge-based workforce, knowledge creation, and creative outputs (such as creative goods and services and online creativity).

6.1 Comparison of IUGP Enablers

A comparison of the IUGP enablers, i.e., governance, regulations, structures and support programs, in US, Norway, Singapore and Qatar is provided in Tables 6.1, 6.2 and 6.3. As one might expect, US has the longest history of IUGP that started as early as 1950s. It is also the only country among the four that has a dedicated industry-university collaboration program (i.e., IUCRC under NSF) since 1970s. R&D and collaboration with industries in both Norway and Singapore have accelerated starting in 1980s mainly driven by: (i) existing level of know-how in certain sectors; (ii) additional government incentives and facilitating regulations; and (iii) the business culture. On the other hand, R&D in Qatar gained attention only recently in late 2000s and early 2010s. Therefore, Qatar is the obviously the youngest nation among the four in terms knowledge infrastructure and innovation ecosystem. Although over the past decade, Qatar has rapidly initiated several programs and projects to boost knowledge creation and transfer in the country, there are number of areas that the country can learn from the other three countries under investigation.

© Springer Nature Switzerland AG 2020

W. Nawaz and M. Koç, *Industry, University and Government Partnerships for the Sustainable Development of Knowledge-Based Society,* Management and Industrial Engineering, https://doi.org/10.1007/978-3-030-26799-5_6

Table 6.1 IUGP related examples of governance and legislation in US, Norway, Singapore, and Qatar

	US	Norway	Singapore	Qatar
Current governing bodies	US Department of Agriculture [1862], National Institutes of Health [1887], National Institute of Standards and Technology [1901], National Science Foundation [1950], Defense Advanced Research Projects Agency [1958], National Aeronautics and Space Administration [1958], Economic Development Administration [1965], US Department of Education [1979], US Department of Energy [1979]	Ministry of Education and Research Norway [1814], Ministry of Trade and Industry Norway [1916], Industrial Development Corporation of Norway [1968], Research Council of Norway [1993], Innovation Norway [2003], Ministry of Health and Care Services Norway [2004]	Ministry of Education Singapore [1960], Economic Development Board [1961], Ministry of Trade and Industry Singapore [1979], SPRING Singapore [1996], Agency for Science, Technology and Research [2001], National Research Foundation [2006]	Ministry of Education and Higher Education [1956], Hamad Medical Corporation [1979], Qatar Foundation [1995], Qatar Development Bank [1997], MEC, Ministry of Transportation and Communication, Ministry of Municipality and Environment; Qatar Research, Development and Innovation (QRDI) Council [2019]
Legislations	Bayh-Dole Act [1980], Stevenson-Wydler Act [1980], Small Business Innovation Development Act [1982], National Cooperative Research Act [1984], Technology Transfer Act [1986], Omnibus Trade and Competitiveness Act [1988], American Technology Pre-eminence Act [1991], National Technology Transfer Improvements Act [1995], Small Business Technology Transfer Act [1992], America COMPETE Act [2007], Leahy-Smith America Invents Act [2011]	Concession Law [1917], Technology or Goodwill Agreement [1979], SkatteFUNN Scheme Warranted by Taxation Law [2002], Law on the Right to Inventions Made by Employees [2003], University and University Colleges Act [2005]	R&D Tax deduction Scheme under Tax Act [1980], Copyright Act [1987], Patents Act [1995], Trademarks Act [1998], Competition Act [2004]	Trademark Law [2002], Law for Protection of Trade Secrets [2005], Law for Establishment of Free Zone for Qatar Science and Technology Park [2005], Patent Law [2008], Law for Establishment of Qatar Development of SMEs [2011]

Table 6.2 IUGP related examples of intermediary structures in US, Norway, Singapore, and Qatar

	US	Norway	Singapore	Qatar
Research Institute/Center Initiatives and Programs with IUGP components	Industry/University Cooperative Research Centers [1973], Engineering Research Centers [1985], Science and Technology Centers [1987], Materials Research Science and Engineering Centers [1994], Focus Center Research Program Centers [1997], Nanoscale Science and Engineering Centers [2001], Nanoelectronics Research Initiative Research Centers [2005], The i6 Challenge: Proof of Concept Centers [2010], National Network for Manufacturing Innovation Institute [2014]	Public Research Institutes [1920], Centers of Excellence [2001], Centers for Research-based Innovation [2007], Centers for Environment-friendly Energy Research [2009], Norsk Katapult Center [2017]	Public Research Institutes, Centers and Consortia [1985], Research Centre of Excellence [2007], Center of Innovation [2007], Campus for Research Excellence and Technological Enterprise [2012]	Qatar Computing Research Institute- QCRI [2010], Qatar Environment and Energy Research Institute-QEERI [2011], Qatar Biomedical Research Institute- QBRI [2012], Interim Translational Research Institute [2014], Medical Research Center-MRC [1998], Sidra Medical and Research Center-SMRC, Qatar Mobility Innovation Center-QMIC [2009], Qatar Bio Bank- QBB [2010], Qatar Innovation Community [2017], Injaz Qatar [2007], Bedaya Center for Entrepreneurship and Career Development [2011], Silatech [2008], Roudha Center for Entrepreneurship [2011]; Qatar Research Development and Innovation-QRDI [2019]

(continued)

Table 6.2 (continued)

	US	Norway	Singapore	Qatar
Clusters	Agricultural Technology Innovation Partnership Cluster [2007], Energy Regional Innovation Cluster [2010], SBA's Pilot Contract based Cluster [2010], Jobs and Innovation Accelerator Clusters JIAC [2011], Advanced Manufacturing-JIAC [2012], Rural JIAC [2012]	Arena Clusters [2002], Norwegian Centers of Expertise [2006], Global Centers of Expertise [2014]	Biopolis [2003], Fushionopolis [2008], Mediapolis [2017]	Education City-EC [1997], Manateq Special Economic Zones [2011]
Science Parks/Business Incubators/TTOs	University-Owned Science Park [1951]/State-Owned Science Park [1952], TTO [early-1980s], Business Incubator [1959]	Science Park [1986], Business Gardens [1998], TTO [2003], Business Incubators [early-2000s]	Science Park [1980], TTO [1992], Blk71 [2011], Blk73 [2015], Blk79 [2015]	Qatar Science and Technology Park-QSTP [2009], QSTP Incubation Center, Qatar Business Incubation Center-QBIC [2014], Digital Incubation Center [2011], NAMA Center [1996], Office of Research Strategy and Impact Management [2014]

Table 6.3 IUGP related examples of the support programs in US, Norway, Singapore, and Qatar

	US	Norway	Singapore	Qatar
Public Procurement	Public Procurement of IC Chips by NASA [early-1960s], Advanced Vehicle and Power Initiative [2010]	Green Public Procurement [2001], Procurement for Innovation [2013]	Ministry of Environment, Ministry of Communication and Information, Singapore Telecoms using Latest Technologies [early-2000s], Crowdsourcing [2017], Spiral Contracting [2017]	Revised Tenders and Bids regulations [2015]
Tax Incentive	The Federal Research and Experimentation (R&E) Tax Credit [1981], Federal Tax-Credits for Solar Systems and Plug-In Electric Vehicles [mid-2000s], State-based Advanced Battery/Technology Collaboration/Solar Tax Credit [mid-2000s]	SkatteFUNN R&D Tax Credit Program [2002]	R&D Tax Incentive for Manufacturing Sector [1980]/Service Sector [1990], R&D Tax Allowance Scheme [2008], R&D Incentive Scheme for Start-up Enterprise [2008], Basic Tax Deduction for R&D [2008], Productivity and Innovation Credit [2010], The Enhanced Tax Deduction for R&D, Super Deduction for R&D [2014]	Qatar Science and Technology Park Tax Free Zone, Tax Free Manateq Special Economic Zones [2011]
Academic Entrepreneurship and Innovation Programs	Academic Entrepreneurship Program [1940]	Academic Entrepreneurship Program [mid-1980s]	Academic Technopreneurship and Innovation Program [2002]	Academic Innovation and Entrepreneurship Program [2012]

(continued)

Table 6.3 (continued)

	US	Norway	Singapore	Qatar
Partnership Programs	Small Business Innovation Research program [1982], Small Business Technology Transfer program [1992], Advanced Technology Program [1990], Technology Innovation Program [2007], Grant Opportunities for Academic Liaison with Industry [1990], Partnerships for Innovation [2000], Seed Fund Support [2009], US Cluster Mapping [2014], Emphasis on Industry Participation in Curriculum Development [2010]	Commercializing R&D Results FORNY [1995], Industrial Research and Development [1994], User-driven Resource based Innovation BIA [2005], Regional R&D and Innovation [2007], Large Scale Programs (FUGE [2002]/BIOTEK2021 [2012], HAVBRUK [2004]/HAVBRUK2 [2016], NANOMAT [2002]/NANO2021 [2012], NORKLIMA [2004]/KLIMAFORSK [2014], PETROMAK [2004]/PETROMAKS2 [2017], RENERGI [2004]/ENERGIX [2013], VERDIKT [2004]/IKTPLUSS [2015]). Start-UP Supports to SMEs [mid-2000s]	Research and Development Assistance Scheme [1981], Technopreneurship21 [1999], Get-UP [2003], Lab-in-RI [2004], Start-up Enterprise Development Scheme SEED [2001], Business Angel Scheme [2005], Early Stage Venture fund [2008], Technology Enterprise Commercialization Scheme [2008], Commercialization of Technology [2006], Exploit Technology Flagship Program[2006], Translational and Clinical Research [2007], Corporate Laboratory@University [2013], Energy National Innovation Challenge NIC [2011], Land and Liveability NIC [2012], NIC Active and Confident Ageing Program [2015], National Cybersecurity R&D program [2013], Marine Science R&D Program [2015], Artificial Intelligence R&D program [2017], Test Bedding and Demonstration of Innovation Research [2015], ASTAR Collaborative Commerce Marketplace [2016], Collaborative Industry Projects [2013], Partnerships for Capability Transformation [2010], Technology Adoption Program [2013], Central Gap Fund [2017]	National Priorities Research Program [2007], Undergraduate Research Experience Program [2006], National Priorities Research Program—Exceptional Proposals [2011], Junior Scientist Research Experience Program [2010], Secondary School Research Experience Program [2010], Graduate Sponsorship Research Award [2013], Postdoctoral Research Award [2015], Conference and Workshop Sponsorship Program [2009], Qatar-UK Research Networking Program [2013], Qatar Innovation Promotion Award [2014], Challenge 22-Innovation Award [2015], Path Towards Personalized Medicine Presentation [2015], Belmont Forum [2015]. Best Representative Image of an Outcome [2015], National Science Research Competition, Academic Health System [2011], Alfikra-Qatar National Business competition [2013], JAHIZ program [2015], SME Equity Program, TASDEER [2011], Qatar Science and Technology Park Funding, Startup and Accelerator Programs, Interactive Programs

Evolution of IUGP-related institutional settings in US, Norway and Singapore indicate that the development of a knowledge-based economy follows a sequential path, starting from building up of operational and adaptive capabilities of local firms, followed by development of a culture and environment to apply innovative solutions through industry relevant research, and lastly with promotion of high tech industries using revolutionary approaches, products or services. A strong and long-term commitment from government, universities and industries is the ultimate driver of innovation. Development of a strong network of local firms mainly the SMEs, and their engagement in technology sharing together with stimulation of private participation in R&D build up a robust foundation for innovation in the country. US, in particular, has been very active in both promoting SMEs and attracting R&D investments from industries. On the other hand, involvement of private sector in R&D in Qatar is insignificant, and in some instances, it is virtually absent. Also, the country appears to be lagging in terms of the size of its local firms, which has an important role in absorbing the wide range of technological developments. This suggests a need to promote local entrepreneurship and private sector participation in R&D in Qatar. Also, US and Singapore have been successful to timely envision the potential new market trends to define their focused sectors. In case of Norway, for example, the Norwegian government emphasized on specific sectors considering the strength of its existing local firms. Similar to the Norwegian setup, the Qatar National Vision 2030 has envisioned four broad pillars in its development strategies. However, there is a need for the government of Qatar to redefine these pillars into more specific focused areas of research and innovation and formulate the country's policies accordingly.

The review of legislative, infrastructural and policy developments in US, Norway and Singapore demonstrates that the best IUGP practices to foster technology development and commercialization involve the following initiatives with a careful, progressive and adaptive planning, implementation, monitoring and continuous improvement: (1) strategic deployment of resources, (2) clustering of innovation actors at regional and national level, (3) development of strong intellectual property regime, (4) promotion of entrepreneurship and high-tech SMEs, and (5) creation of market demands for innovative products and services.

US is known to be leading example in the IUGP practices along with Germany, which is not included in this study. IUCRCs, US Science Parks, Bayh-Dole Act of 1981, SBIR represent some of the best IUGP practices organically initiated, implemented and improved in the US over time. Similarly, the timely and strategic decisions made by Norwegian government to deploy its natural resources in a way which supports the technological competence of the country is an important lesson for the hydrocarbon-based economies. Also, Singapore's Science Parks and Clusters, and its entrepreneurship programs are also exemplary practices for emerging nations, such as Qatar.

6.2 Comparison of GII and Indicators

After developing a deep theoretical understanding of the IUGP mechanism in the four countries, the discussion can essentially benefit from the quantitative assessment of the effectiveness of IUGP settings in these countries. Since one of the core objectives and outcome of IUGP is promotion of innovation in the country, the 'innovative-related performance and capacity' of a country can be a good measure to gauge the progress and effectiveness of IUGPs. In this section, we compare the four countries over quantitative measures of GII.

The GII, published since 2007, is an annual country-based ranking of performance and capacity for innovation. The index captures the multi-dimensional indicators of innovation which can benefit long-term output growth and improved productivity. The evolving framework of GII is shown in Fig. 6.1. At the top level, GII can be subdivided into two sub-indexes, i.e., innovation input and innovation output (Cornell University, INSEAD, and WIPO 2017). The input and output sub-indexes are divided into input and output pillars, respectively; 5 pillars for innovation input and 2 pillars for innovation output. Each of the pillar is further divided into three sub-pillars, which are measured for each country through a set of quantifiable indicators. The total number of indicators used in 2017s ranking is 81, however, not all of the indicators have equal contribution towards the calculation of the sub-pillar score; the ones with half weight contribution are marked as '(a)' in Fig. 6.1. Also, in some cases the indicators are absolute values (hard data) obtained from competent sources, such as statistics published by United Nations, while in other instances, these may be composite indicators, or indicators based on survey results. More details on the indicator calculation and estimation can be obtained from the 2017s GII report (Cornell University, INSEAD, and WIPO 2017).

The comparison of the GII and its sub-indexes (input and output) for the four countries under discussion is presented in Fig. 6.2. Qatar ranks lowest in comparison to the other three countries in innovation input and output category, and subsequently in the overall innovation index. However, the difference between the innovation input and output for Qatar is lower than that of Norway and Singapore. This is a positive sign for the gas-rich country as it shows better efficiency of conversion of inputs to the outputs compared to the other two nations. At the same time, this also implies that Qatar does have the potential to be among the leading innovative countries in the world, however, there may be some certain domains where the country needs to do better. We will explore these domains by looking closely into the indicators of GII.

In order to identify the potential improvement opportunities for Qatar, we analyzed the performance of these countries on the GII indicators' level. A systematic methodology was developed to highlight the most impactful indicators from Qatar's perspective; when the sum of the differences of Qatar's sub-pillar score and other three countries' sub-pillar score equals or increases 75, all indicators in that sub-pillar are considered for the analysis. For example, the sum of the difference of Qatar's score from other countries in the sub-pillar 'political environment' is 35.2 (<75), as

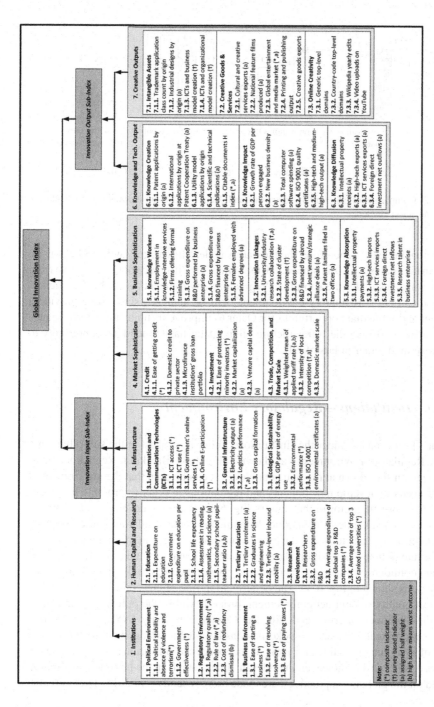

Fig. 6.1 Framework of GII. Adopted from Cornell University, INSEAD, and WIPO (2017)

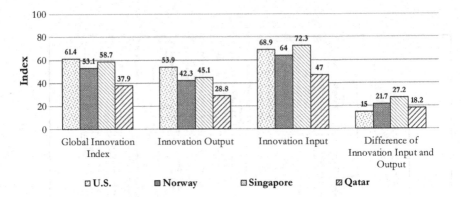

Fig. 6.2 Comparison of US, Norway, Singapore, and Qatar over GII, sub-indexes (input and output), and innovation efficiency

shown in Table 6.4. Therefore, the indicators under this category will not be considered in the analysis. On the other hand, the sum of the difference of Qatar's score from other countries under the sub-pillar 'regulatory environment' is 85.3 (>75), hence, the indicators of 'regulatory environment' will be analyzed since these provide a greater and immediate opportunity for Qatar to improve its innovation performance and capacity. Beside the Infrastructure pillar, Qatar has tremendous opportunity to improve in all other pillars of GII.

6.2.1 Institutions—Regulatory Environment

Qatar's performance under the 'Institutions' pillar is weakest in the sub-pillar category 'Regulatory Environment', as shown in Fig. 6.3a. Breaking 'Regulatory Environment' further down at the indicators level, it can be seen that the government of Qatar needs to develop new policies and regulations in order to promote the development of private sector in the country Fig. 6.3b. Since the development of private sector is the backbone of economic diversification, the regulating bodies in Qatar should design means to assess the effectiveness and the extent of implementation of the current policies. In addition, the government of Qatar must take necessary actions to gain confidence of the global community when it comes to the rule of law in the country, as shown in Fig. 6.3c. The betterment in rule of law comes with the betterment of quality of contract enforcement, property rights, the police and the courts, and the likelihood of crime and violence (Cornell University, INSEAD, and WIPO 2017). Admittedly, the state's spokespersons have acknowledged the importance of rule of law, human rights, and judicial independence on various occasions (Qatar News Agency 2017), however, the relevant amendments in the law and constitution, based on the global best practices, are made more recently—in 2015. Therefore, the

Table 6.4 Identification of crucial innovation indicators for Qatar

Pillar	Sub-pillars	Score of countries in sub-pillars				Sum of the difference of the scores $(A - D) + (B - D) + (C - D)$
		US (A)	Norway (B)	Singapore (C)	Qatar (D)	
Institutions	Political environment	80.3	90.8	96.9	77.6	35.2
	Regulatory environment	90.4	94.9	98.6	66.2	85.3
	Business environment	88.1	89.6	87.6	74.6	41.5
Human capital and research	Education	54.7	64.3	44	37.2	51.4
	Tertiary education	38.1	40.2	80.5	55.7	−8.3
	Research and development	78.8	55.5	66.5	7	179.8
Infrastructure	ICTs	85.2	80.9	87.8	68.5	48.4
	General infrastructure	52.8	72.4	57.7	67.6	−19.9
	Ecological sustainability	45	54.7	62	38.3	46.8
Market sophistication	Credit	85.5	55.2	63.4	28.6	118.3
	Investment	72.2	46.3	75	30.1	103.2
	Trade, competition, market scale	92.7	70.2	75.2	68.9	31.4
Business sophistication	Knowledge workers	67.4	67.7	73.2	20.4	147.1
	Innovation linkages	46.6	40	46.5	33	34.1
	Knowledge absorption	55.2	37.1	68.9	30.6	69.4

(continued)

Table 6.4 (continued)

Pillar	Sub-pillars	Score of countries in sub-pillars				Sum of the difference of the scores $(A - D) + (B - D) + (C - D)$
		US (A)	Norway (B)	Singapore (C)	Qatar (D)	
Knowledge and technology outputs	Knowledge creation	63.4	34	27.7	3.5	114.6
	Knowledge impact	52.5	44.1	47.2	33.9	42.1
	Knowledge diffusion	47.3	34.4	67.1	32	52.8
Creative outputs	Intangible assets	50.1	51.5	49	51.8	−4.8
	Creative goods and services	48.2	27.4	36	12.2	75
	Online creativity	65.4	57.8	37.8	22.4	93.8

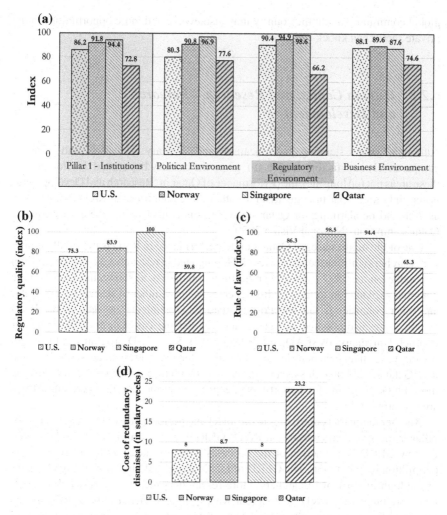

Fig. 6.3 Comparison of the innovation performance of US, Norway, Singapore, and Qatar over: **a** 'Institutions' pillar and its sub-pillars; and the indicators of the crucial sub-pillar 'Regulatory Environment', including, **b** regulatory quality, **c** rule of law, and **d** cost of redundancy dismissal

outcomes of Qatar's efforts to uphold rule of law as per the international standards are yet to be observed.

Finally, the country can also benefit significantly from improving the regulations pertaining to employment conditions, as shown in Fig. 6.3d, especially related to hiring, working hours, advance notice requirements, and severance payments due when terminating a redundant worker (Cornell University, INSEAD, and WIPO 2017). By bringing the employment regulations to an international standard, not only that Qatar can be a home to the best talent in the world, the perception of the global community about the country may also evolve and more opportunities in the private sector can knock the doors.

6.2.2   Human Capital and Research—Research and Development

Although Qatar's standing in 'Education' and 'Tertiary Education' sub-pillars is comparatively satisfactory (Fig. 6.4a), given the country has started development in these areas not too long ago, the performance of Qatar in 'Research and Development' is certainly under-par in comparison to other three countries; an R&D index as low as 7 should be alarming for Qatar since R&D is considered as the cornerstone of Qatar's ambitious National Vision 2030.

One of the main causes of Qatar's underperformance in R&D is the small number of researchers (fulltime equivalence—FTE) engaged in research activities per million of population, i.e., 597.1 researchers/million population, as shown in Fig. 6.4b. One potential explanation to the big difference between number of researchers in Qatar and other countries is that the Postgraduate Ph.D. students engaged in the R&D activities are also counted as 'researchers' as per the calculation of the indicator (Cornell University, INSEAD, and WIPO 2017). Since PhD programs in Qatar are (by far) fewer than the countries in comparison, the overall researchers' count is also less. Qatar should take this as an opportunity to enhance its workforce and develop more Ph.D. programs to support QNV, especially in the setting where eight IBCs are operating in EC.

Another factor that contributes to the under-par performance of Qatar in R&D sub-pillar is the gross expenditure on R&D (GERD), as shown in Fig. 6.4c. However, the low GERD is a function of the previous indicator, i.e., number of researchers per million of population. The GERD of US, Norway and Singapore is high because of the high number of researchers; more researchers mean more research projects and more budget for R&D. In a similar way, Qatar can increase its GERD through the oil and gas revenue as a mean to attract more researchers—otherwise, both of these numbers, researchers and GERD, may remain lower in comparison to other countries.

The third underachieving indicator in the R&D sub-pillar is the average expenditure of top 3 global companies on R&D (Fig. 6.4d). Since none of Qatar's companies are listed in the European Union Industrial R&D Investment Scoreboard, from where this indicator is obtained, the country scored a zero on this indicator, which significantly dropped its overall performance in the R&D sub-pillar. The top five companies in this list, i.e., with most R&D funding in the fiscal year 2015–16, are Volkswagen, Samsung, Intel, Alphabet, and Microsoft (Joint Research Center 2016)—three of which are based in the US.

With the presence of only one university in the QS ranking, i.e., QU, Qatar's score in QS university ranking is least among the countries in comparison, as shown

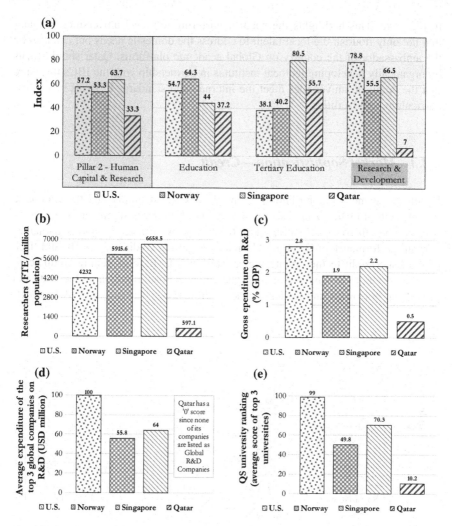

Fig. 6.4 Comparison of the innovation performance of US, Norway, Singapore, and Qatar over: **a** 'Human Capital & Research' pillar and its sub-pillars; and the indicators of the crucial sub-pillar 'Research & Development', including, **b** researchers, **c** gross expenditure on R&D, **d** average expenditure of top three global R&D companies, and **e** average score of top three universities as per QS ranking

in Fig. 6.4e. This highlights the need of nurturing top-level universities in Qatar that not only flourish the local talent to address the domestic needs but can also be the ambassador of the country on Global academic platforms. Qatar should focus on organically developing its local institutes in partnership with the IBCs to ensure that the domestic universities meet the international standards in terms of faculty, curriculum, and facilities.

6.2.3 Market Sophistication—Credit

Despite the generous contributions from Qatar's Royal family and the economic involvement of QDB, Qatar's score on 'Credit' sub-pillar is significantly low compared to the other three countries (Fig. 6.5a). The two causes of the unsatisfactory 'Credit' performance of Qatar are 'difficulty of getting credit' (Fig. 6.5b) and 'lack of domestic credit to the private sector' (Fig. 6.5c). 'Ease of getting credit' captures the extent to which the monetary guarantee and bankruptcy laws facilitate and protect the borrowers and lenders in the credit process. The indicator also reflects

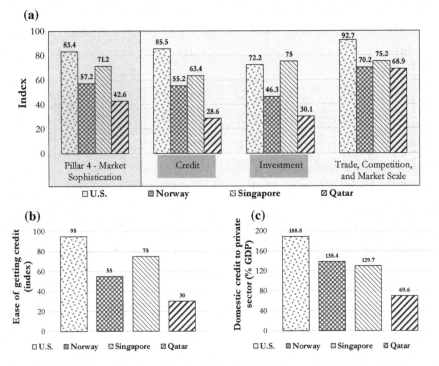

Fig. 6.5 Comparison of the innovation performance of US, Norway, Singapore, and Qatar over: **a** 'Market Sophistication' pillar and its sub-pillars; and the indicators of the crucial sub-pillar 'Credit', including, **b** ease of getting credit, and **c** domestic credit to private sector

on the countries' rules and practices related to coverage, scope, and accessibility of credit information (Cornell University, INSEAD, and WIPO 2017). Although Qatar has taken a step forward through the amendments in the public procurement law, i.e., Law No. 26 of 2005, which aims to support SMEs by waiving the performance bonds and guarantees for public procurement, the government should further focus on: (1) encouraging independent financing schemes under legal framework to ensure secured private leasing; and (2) establishing effective credit registries and bureaus to facilitate the financing schemes (Doing Business 2016). A well-established crediting mechanism, as can be seen in other countries (Fig. 6.5b), would significantly improve the numbers and quality of SMEs and private leasing companies in the country, which would reduce the financial burden on the government.

Qatar also under performs in the 'domestic credit to private sector' (Fig. 6.5c). The comparatively low score of Qatar in this indicator means that the financial resources, e.g., loans, non-equity securities, and trade credits, available to support the private sector in Qatar through financial institutions, e.g., banks and financial corporations, are low as compared to other countries. One of the potential reasons for this low score may be the lack of financial institutions in Qatar, which may include monetary authorities, deposit and savings money banks, finance and leasing companies, money lenders, insurance corporations, pension funds, and foreign exchange companies (Cornell University, INSEAD, and WIPO 2017). An increase in the number of such institutions would automatically increase the domestic credit available to the private sector.

It is important to note here that there is another indicator for calculating 'credit' sub-pillar, i.e., microfinance gross loans, but since this indicator is not applicable to any of the countries (missing data) under discussion, it will not be discussed. It should also be noted that the indicators which are not applicable to a country, due to missing data, are not considered in the calculation of sub-pillar score (Cornell University, INSEAD, and WIPO 2017).

6.2.4 Market Sophistication—Investment

Qatar's performance in the 'Investment' sub-pillar is also below average in comparison to the other three countries (Fig. 6.5a). The indicators of 'Investment' sub-pillar are shown in Fig. 6.6. From a market's perspective, the investors generally tend to invest in countries where their investments are secured—to ensure risk free profit. The first indicator of investment sub-pillar, 'ease of protecting minority investors', captures the quality and effectiveness of the conflict of interest regulation and the extent of shareholder governance in a country (Fig. 6.6a). The indicator is calculated based on a combination of information pertaining to the extent of disclosure index, extent of director liability index, access to legal expenses information, extent of shareholder rights index, extent of ownership and control index and extent of corporate transparency index (Cornell University, INSEAD, and WIPO 2017). Qatar needs to re-evaluate its investors' safeguard measures since with the present score

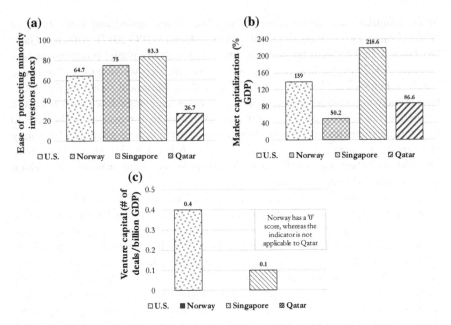

Fig. 6.6 Comparison of the innovation performance of US, Norway, Singapore, and Qatar over the indicators of the crucial sub-pillar 'Investment', including **a** ease of protecting minority investors, **b** market capitalization, and **c** venture capital

of 26.7 the foreign investors, who are a key to private sector development, are less likely to be interested in investing in the country. The overhauling of the business regulations can be done based on Singapore's model of corporate governance, which is consistently ranked amongst the best in the world (Lin and Ewing-Chow 2014).

Although Qatar's performance on market capitalization is better than that of Norway (Fig. 6.6b), development in private sector, positive environment for foreign investment and an increase in the number of listed companies can make this score comparable to that of US and Singapore. In this regard, the country will significantly benefit from the new law on PPPs (Table 5.1), which is expected to provide investment opportunities of up to US $20 billion in variety of sectors. In order to get the maximum benefit from this opportunity, the government should start to devise plans for investors ahead of time to ensure a wide foreign participation in the economic diversification and development of Qatar.

The last indicator of 'Investment' sub-pillar is the number of venture capital deals per billion of GDP (purchasing power parity). The data for venture capital deals is not available for Qatar which is why the indicator is not applicable to the country (Fig. 6.6c). It is an opportunity for the relevant agencies and ministries in the country to ensure the availability of such data since beside establishing comparisons and doing benchmarking, publishing such data portrays an investor-friendly image of the country on the global platform. MEC should take appropriate steps to develop mechanisms to measure and publish the investment related data.

6.2.5 Business Sophistication—Knowledge Workers

Similar to 'Market Sophistication' pillar, Qatar's performance in the 'Business Sophistication' pillar is under-par, especially in the 'Knowledge Workers' sub-pillar, as shown in Fig. 6.7a. As seen earlier, Qatar's score on 'researchers engaged in research activities per million of population' is relatively low in comparison to the

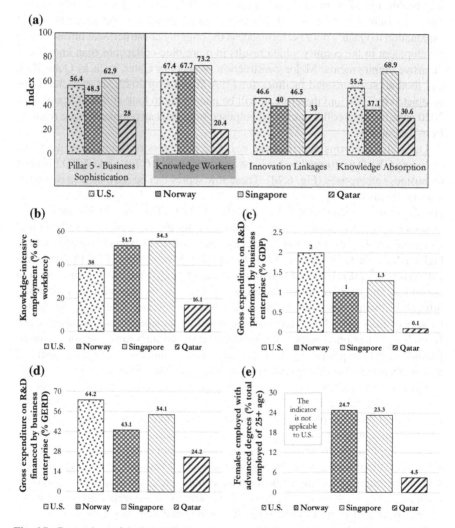

Fig. 6.7 Comparison of the innovation performance of US, Norway, Singapore, and Qatar over: **a** 'Business Sophistication' pillar and its sub-pillars; and the indicators of the crucial sub-pillar 'Knowledge Workers'[2], including **b** knowledge-intensive employment, **c** GERD performed by business enterprises, **d** GERD financed by business enterprises, and **e** Females employed with advanced degrees

other three countries (Fig. 6.4b), the same trend continues for the 'knowledge intensive employment' (Fig. 6.7b). There are three employment categories that fall under knowledge intensive employment (International Labour Office 2012): (1) Managers (e.g., chief executives, managing directors, legislators, commercial and administrative managers, production and services managers, and others); (2) Professionals (e.g., science and engineering, health, teaching, business and operations, ICT, legal and social, and others); and (3) Technicians and Associates (e.g., science and engineering, health, teaching, business and operations, ICT, legal and social, and others). Since the indicator is measured as the percentage of total workforce, a potential explanation to Qatar's poor performance is the ongoing constructional infrastructure development in the country which results in more blue-collar jobs than knowledge intensive employments. Major construction projects in Qatar, such as Qatar Rail, and football stadiums and facilities for FIFA World Cup 2022, are expected to be completed by 2020 and therefore, it will be interesting to look at this indicator after 2020 once the blue-collar workers (composed of expatriate population) are released from the country.

Qatar also performs poorly in the next two indicators of 'knowledge workers', including GERD performed by business enterprises (Fig. 6.7c), and GERD financed by business enterprises (Fig. 6.7d). These low scores can be attributed to two common problems that are realized from the discussion thus far: (1) small size of the private business sector in the country; and (2) lack of interest of business enterprises in R&D. According to United Nations Educational, Scientific and Cultural Organization (UNESCO), Qatar's primary source of R&D expenditure comes from higher education departments (36.6%) followed by government (31.2%) and business enterprises (24.2%) (UNESCO Institute for Statistics 2012). However, since most of the entities in Qatar are either funded or supported by the government, it becomes difficult to distinguish between the exact source of funding. Therefore, firstly, the sources of fund should be clearly differentiated by the education departments, and secondly, as mentioned before, the development of private business sector should be handled as the top priority by the leadership of the country—so that the investment and involvement of business enterprises in R&D also increase.

Qatar's performance on 'female employed with advanced degree' is also not promising (Fig. 6.7e). However, it has the same explanation as 'knowledge intensive employment' (Fig. 6.7b); since the indicator is measured as the 'percentage of total employment', high number of blue-collar jobs in construction sector outnumbers the females employed with advanced degree. The indicator will be more interesting to study after the infrastructure development in Qatar phase-out, especially since the number of female university students in the country are about double the number of male university students and the females are 60% of all university graduates in the country (Qatar Ministry of Development Planning and Statistics 2014).

It is important to note that while 'Knowledge Absorption' sub-pillar (Fig. 6.7a) did not pass the screening criteria (Table 6.4) of the analysis of crucial indicators, Qatar's performance in this sub-pillar is certainly not up to the mark. The country needs to find ways to increase the following in order to absorb more knowledge in its ecosystem: IP payments; FDI net inflows; and research talent in business enterprises.

At the same time, the country should decrease the following for more knowledge absorption: high-tech imports and import of ICT services.

6.2.6 Knowledge and Technology Outputs—Knowledge Creation

'Knowledge and Technology Outputs' is one of the only two output pillars of GII. Qatar's performance in 'Knowledge Creation' sub-pillar of 'Knowledge and Technology Outputs' is poor in comparison to other three countries, as shown in Fig. 6.8a. All indicators of 'Knowledge Creation' are extremely important from Qatar's perspective since the country is doing poor on all of them.

Qatar has received a low score of 0.13 on the number of resident patent applications filed at a given national or regional patent office (Fig. 6.8b). The country has also received a low score on the number of international patent applications filed by residents at the Patent Cooperation Treaty in 2017 (Fig. 6.8c). These statistics are alarming for Qatar since these were even lower in 2016 and 2015 (Cornell University, INSEAD, and WIPO 2016). A continuous poor performance of the country on patents' indicators shows a lack of focus on translation of fundamental research into commercializable commodity. This also raises several questions on the quality of work being carried out in the research institutions and incubation centers in Qatar.

Although Qatar expects to be a research, innovation and knowledge hub in the region, the results of scientific and technical publications (Fig. 6.8d), and citable documents H-index (Fig. 6.8e) make the situation further uncollectible for the country. Nevertheless, since knowledge ecosystem in Qatar is comparatively still under development stage, there may be a significant improvement in this indicator in the coming decade. Meanwhile, the country should invest in more research-focused institutions than instructional institutions. Also, the researchers should not only focus on increasing their number of scientific and technical publications but also on improving the quality of the research work in order to publish the research findings in well reputed journals of higher impact. This will improve the standing of the country on 'citable documents H-index'.

6.2.7 Creative Outputs—Creative Goods and Services

'Creative Goods and Services' sub-pillar is one of the reasons for Qatar's (comparatively) poor performance in 'Creative Outputs' pillar of GII, as shown in Fig. 6.9a. The sub-pillar is divided into five indicators; however, since Qatar's data is missing on two indicators, only three indicators will be discussed. Nevertheless, the country should make efforts in the future to publish yearly data of the missing indicators

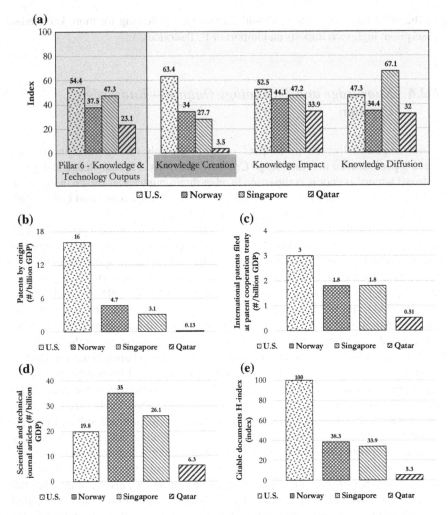

Fig. 6.8 Comparison of the innovation performance of US, Norway, Singapore, and Qatar over: **a** 'Knowledge and Technology Outputs' pillar and its sub-pillars; and the indicators of the crucial sub-pillar 'Knowledge Creation'[4], including **b** Patents by origin, **c** international patents filed at PCT, **d** scientific and technical journal articles, and **e** citable documents H-index

which include 'cultural and creative services exports (as percentage of total trade)' and 'national feature films produced (as number of films/million population)'.

As shown in Fig. 6.9b, Qatar's score on 'global entertainment and media market' is significantly lower than that of US, Norway, and Singapore. This indicator captures five-year forecast and five-year historic consumer and advertiser spending data (Cornell University, INSEAD, and WIPO 2017). It covers sectors that invest and operate in the entertainment and media industry such as book publishing, filmed entertainment, internet advertising, magazine publishing, music, newspaper publish-

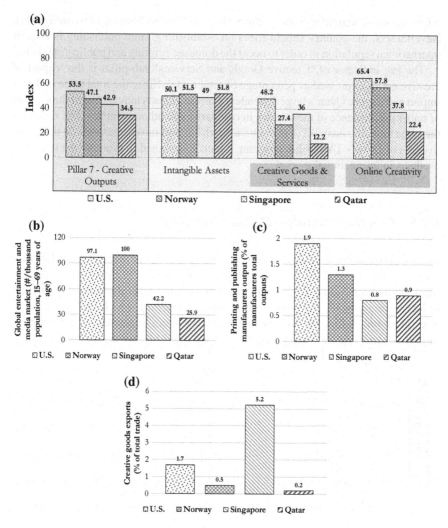

Fig. 6.9 Comparison of the innovation performance of US, Norway, Singapore, and Qatar over: **a** 'Creative Outputs' pillar and its sub-pillars; and the indicators of the crucial sub-pillar 'Knowledge Creation', including **b** global entertainment and media market, **c** printing and publishing manufacturers output, and **d** creative goods exports

ing, out of home advertising, TV advertising, video games and others. Qatar can use the FIFA 2022 World Cup (to be held in the country) as a branding strategy to support its entertainment and media industry (Ginesta and de San Eugenio 2016).

Qatar's performance on 'printing and publishing output' is relatively better than other indicators in 'Creative Goods and Services' sub-pillar (Fig. 6.9c). The 'printing and publishing output' indicator captures the economic activities carried out in publishing, printing and reproduction of recorded media as a percentage of all other economic activities in the country (United Nations Statistics Division 2008). Nevertheless, the country should focus on establishing more publishing houses of international reputation in order to boost the domestic printing and publication sector.

The last indicator of 'Creative Goods and Services' sub-pillar is the 'export of creative goods expressed as percentage of total trade'. Since the total trade includes import and export, Qatar's large dependence on the imports may be one of the reasons of under performance of the country in comparison to other countries (Fig. 6.9d). At the same time, the mere 0.2% of creative goods' export raises questions on the effectiveness of QDB's TASDEER program, which is designed to promote the domestic products in international market.

6.2.8 Creative Outputs—Online Creativity

Qatar's performance is relatively poor in 'Online Creativity' sub-pillar of 'Creative Outputs' (Fig. 6.9a). The first two indicators of 'Online Creativity', i.e., 'generic top-level domains' (Fig. 6.10a) and 'country-code top-level domains'(Fig. 6.10b), reflect on the country's performance in internet learning which includes hosting generic (.com, .info, .net, and .org) and country specific domains (.qa for Qatar, .no for Norway, .sg for Singapore, and .us for United States). As the indicators are

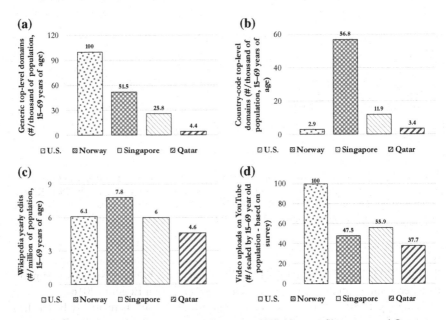

Fig. 6.10 Comparison of the innovation performance of US, Norway, Singapore, and Qatar over the indicators of the crucial sub-pillar 'Online Creativity', including **a** generic top-level domains, **b** country-code top-level domains, **c** wikipedia yearly edits, and **d** videos uploaded on YouTube

measured by 'number of domains per thousand of population between 15–69 ages', Qatar's performance was expected to be better than its current standing—given Qatar has lowest population among the countries in comparison. The low number of generic and country specific domains show a low presence of Qatari firms on the internet; these numbers automatically increase if the country has more business enterprises and SMEs which use internet to be more visible to the domestic and international community.

Qatar's performance is also weak on the other two indicators of 'Creative Outputs', i.e., 'Wikipedia yearly edits' (Fig. 6.10c) and 'Video uploaded on YouTube' (Fig. 6.10d). As mentioned before, the country should have performed better on these two indicators since these are also measured as a function of total population, and Qatar's population is lowest in comparison to other countries. The relatively low numbers on both indicators show a lack of interest of Qatar's population in the online creative activities, which in the modern world are also income generating platforms—such as YouTube. The country needs to promote and encourage the participation of its people, especially the young generation, in the online creativity in order to develop a culture of research and innovation. Propagation of creativity at an early stage of development of young generation will help in producing the human network which takes the knowledge intensive route to economic development.

References

Cornell University, INSEAD, & WIPO. (2016). *The global innovation index 2016: Winning with global innovation*. Ithaca, Fontainebleau, and Geneva.

Cornell University, INSEAD, & WIPO. (2017). *The global innovation index 2017—innovation feeding the world*. Ithaca, Fontainebleau, and Geneva.

Doing Business. (2016). *Measuring business regulations*. http://www.doingbusiness.org/data/exploretopics/getting-credit/what-measured.

Ginesta, X., de San Eugenio, J. (2016). The use of football as a country branding strategy. Case study: Qatar and the catalan sports press. *Communication & Sport, 2*(3), 225–241.

International Labour Office. (2012). *International standard classification of occupations*. Geneva. http://www.ilo.org/public/english/bureau/stat/isco/docs/publication08.pdf.

Joint Research Center. (2016). Economics of industrial research and innovation—the scoreboard. http://iri.jrc.ec.europa.eu/home#close.

Lin, L., Ewing-Chow, M. (2014). *The doing business indicators in minority investor protection: The case of singapore*. http://law.nus.edu.sg/wps/pdfs/007_2014__LinLin.pdf.

Qatar Ministry of Development Planning and Statistics. (2014). *Qatar social statistics 2003–2012*. Doha. http://www.mdps.gov.qa/en/knowledge/Publications/Social/Soc_Qatar_Social_Statistics_En_2014.pdf.

Qatar News Agency. (2017). Qatar vows to uphold the principle of rule of law. *Gulf Times*. http://www2.gulf-times.com/story/566362/Qatar-vows-to-uphold-the-principle-of-rule-of-law.

UNESCO Institute for Statistics. (2012). Science, technology and innovation: GERD by source of funds. http://data.uis.unesco.org/#.

United Nations Statistics Division. (2008). *International standard industrial classification of all economic activities—Rev04*. New York.

Chapter 7
A Survey on the Current Status and Future of IUGPs in Qatar: Challenges, Opportunities, and Recommendations

Abstract Qatar's vision for innovation-driven knowledge economy and sustainable development has been the key motivation for its leadership to strategize and invest in the knowledge ecosystem. Although the outlook of knowledge infrastructure in Qatar, including national and international universities, research and incubation centers, science park, and technology transfer offices, resembles that of the United States, Singapore, and Norway, the country is yet to achieve the same level of excellence in terms of the outcomes. The primary reason for the lack of outputs is that the concept of economic transformation is still new for Qatar and its citizens, and since the overall system is at an infancy stage, some crucial elements of Industry-University-Government partnerships are missing at the execution level. We have already discussed the areas of concern for Qatar in the previous chapter in the light of the Global Innovation Index (Chap. 6), which include regulatory environment, research and development, market and business sophistication, and creativity. In order to learn the reasons behind these shortcomings–albeit the country's commitment to transform to a knowledge-based economy–we conducted interviews and survey with the experts that have firsthand experience of working in the R&D and Industry-University-Government settings in Qatar. In the first part of this chapter, we present the findings of the interviews and survey while in the second part, we propose recommendations to make the knowledge ecosystem in Qatar more effective and efficient.

7.1 Background

The role of IUGPs has been increasingly realized by countries across the world as a mean of economic diversification and development (Kaklauskas et al. 2018). Qatar is a country in transition from a resource-rich to knowledge-based economy. Thus, with its emerging innovation system and quickly expanding R&D infrastructure, it has an increasing need to promote close, effective, and tailored collaboration mechanism between industry, academia and government in order to create an environment that supports economically relevant innovation, rapid technology transfer, and a diversified economy with high-value added sectors and products.

© Springer Nature Switzerland AG 2020
W. Nawaz and M. Koç, *Industry, University and Government Partnerships for the Sustainable Development of Knowledge-Based Society,*
Management and Industrial Engineering, https://doi.org/10.1007/978-3-030-26799-5_7

There is a common perception that while the major players in the academic cor-
ridor, such as QU and QF and its academic members clustered in EC, seem to be
ready to play their part (Qatar News Agency 2018; Qatar University 2018), it is the
industry (i.e., mainly the oil and gas sector), intermediaries, and government which
often lack preparedness in playing their role (Ahmed 2018). Also, since most of
the entities operating in Qatar are owned, funded, or supported by the government
(directly or indirectly), it becomes difficult to understand the true nature and not-so-
well working mechanism of the existing IUGPs (Gremm et al. 2018). However, the
above statement is merely speculation; there is a lack of empirical evidence to support this claim. Notwithstanding, it is still unclear how organizations currently form
partnerships with one another under one umbrella (government) and how different
players benefit from the outcomes of these partnerships. In order to understand the
complex interactions of IUGP players in Qatar, we decided to reach out the experts
in this field with the firsthand experience to:

- understand the current status of IUGP settings;
- learn the weaknesses and strengths of IUGP mechanism(s);
- compare the IUGP settings in the country with other countries; and
- develop and provide tailored policy recommendations to improve the overall system.

In general, we realized during the sample selection phase that there has not been
much awareness about the IUGPs in Qatar and the necessity of industry-university
collaborations for economic diversification. There are currently very limited number
of personnel who are working towards creating partnerships and finding opportunities
for different entities to work together.

7.2   Design and Validation of the Survey

Following a comprehensive review of literature and limited interviews with key
players in academia and industry, an instrument (survey) was designed to collect
IUGP-related information from the relevant actors in Qatar. The survey was divided
into four parts to reflect on the following:

- Awareness and understanding of IUGP mechanism and outcomes;
- Position of IUGPs in Qatar and its weaknesses compared to other countries;
- Level of satisfaction of the respondents with regards to the current IUGP status of
  the country; and
- Recommendations for improving the IUGPs.

Various sources, including academic literature and national IUGP reports of countries, such as US, Germany, Norway, Singapore, Canada and Indonesia, were con-
sulted before conceptualizing the survey and preparing the first draft (Board of Trade
of Metropolitan Montreal 2011; Edmondson et al. 2012; Hall 2004; Moeliodihardjo
et al. 2013). Nevertheless, since Qatar's settings are different from other parts of the

world, IUGP literature relevant to Qatar was reviewed and reflected in the survey to ensure the relevance of survey items with the context of the country (Ahmed 2018; Conventz et al. 2015; Gremm et al. 2018; Weber 2014). Consequently, the important variables collected from the literature and reports are as follows:

- Commitment from the leadership
- Shared vision, coordination, and synergy
- Culture and dialogue
- Mutual trust
- Balancing individual and institutional interest
- Support, funding, and incentives
- Interdisciplinary strategy
- Long-term partnership
- Research and learning
- Innovation-oriented initiatives
- Human resources
- In-house facilities
- Flexibility and relevance of the partnerships
- Redefining the role of universities—problem-solving institutions for the society
- Social value creation through partnerships
- Flexible administrative control
- Importance of outcomes (but not an overemphasis).

Furthermore, in order to ensure the relevance of survey items for specific IUGP sectors, the survey was conducted in four distinct sections for industries, academia, government, and intermediaries. The intermediaries were added as a separate section to create a distinction between the primary IUGP institutions and supporting and facilitating institutions. Most of the questions were the same in all four sections— only a few questions were sector-specific; following the demographics and profiling and general questions about IUGP in Qatar, the respondents were redirected to the survey sections specific to their sector.

While it was difficult to distinguish between the jurisdiction of many organizations since most of these are owned, funded, or supported by the government (directly or indirectly), we developed a guideline to draw the difference between the types of organizations; it was decided to limit the government sector to the ministries and councils, and other major players, such as QP and HMC, which are although operated under the government umbrella, are assumed independent. This assumption may be considered as one of the major limitations of this work, but there was no other possible way to examine the interaction of IUGP actors in Qatar due to the government-dominant structure of organizations. Some examples to draw the sector-wise difference between various entities are provided in Table 7.1.

The first draft of the questionnaire consisted of a combination of multiple choice, multiple answers, five-point Likert scale, and open-ended questions. The total number

Table 7.1 Sector-wise classification of organizations for survey

Sector	Examples
Industry	The industry sector covers some government-owned and all private and non-profit organizations producing and marketing goods and services. It also includes their affiliated institutes and centers. A few examples include Qatar Petroleum, Qatar Steel, Qatar Airways, Ooredoo, Al Jazeera, Kahramaa, Hamad Medical Corp, SIDRA, IBM Qatar, Microsoft Qatar, Qatar Shell Research Technology Center, ExxonMobil Research Center, Total, ConocoPhillips, GE, Chevron, Ali bin Ali, Mall of Qatar and others
Academia	The academic sector includes all Higher Education Institutes in Qatar, and their affiliated research institutes and centers. A few examples include Qatar University, Hamad Bin Khalifa University, QEERI, QCRI, QBRI, Doha Institute for Graduate Studies, Community College Qatar, Qatar Aeronautical College, Education city branch campuses like Texas A&M University at Qatar and others
Government	The government sector includes all public authorities with governance roles. It also includes their affiliated institutes and centers. A few examples include Ministry of Economy and Commerce, Ministry of Education and Higher Education, Ministry of Municipality and Environment, Ministry of Transportation and Communication, Supreme Council of Health, Supreme Council of Delivery and Legacy, Qatar Foundation, Qatar National Research Fund, Qatar BioBank, Qatar Foundation Research & Development and others
Intermediary	Organizations that facilitate the innovation and entrepreneurship activities in the country. Intermediaries work with industry, university and/or government to support innovation by translating fundamental research. A few examples include Qatar Science and Technology Park, Qatar Development Bank, Qatar Mobility and Innovation Center, Qatar Business Incubation Center, Silatech, INJAZ Qatar, Bedaya Center for Entrepreneurship and Career Development, Roudha Center for Entrepreneurship and others

of the questions were 95.[1] In order to validate the questionnaire, first, one-on-one interviews were conducted with experts from industry, business, university, government, and intermediary circles in Qatar. In total, 18 experts participated in the interviews (4 from industry, 7 from universities, 4 from government, and 3 from intermediaries). The selection of interviewees was not based on the representativeness of the total population, rather these professionals were chosen due to their involvement and unique role in the IUGP settings of Qatar.

Based on the suggestions of the interviewees, one question was removed from the government-specific part and eight questions were added to the final questionnaire, which consisted of 102 questions. The results of the interviews will be discussed later in this chapter along with the results of the final survey.

---

[1] The number corresponds to the sum of all questions. In reality, the respondents had to answer only those questions which were relevant to their sector. Roughly, there were around 38 questions which each respondent had to answer.

7.2.1 Selection of Respondents

During the interviews, snowball approach (chain referral sampling) was used to increase the pool of experts for the final survey (Biernacki and Waldorf 1981); at the end of each interview, the interviewees were asked to identify relevant people who they think are pertinent for this study and will be interested in participating in the survey. As mentioned before, the approach used in this study for the selection of potential respondents is 'expert sampling approach', where informants are nonrandomly selected based on their relevance to the objective of the study and not based on the representation of the overall population (MacDonald et al. 2019). Through the chain referrals and personal contacts, the pool of experts to be contacted to participate in the survey reached 803 by the end of the interview phase (241 from industry, 350 from academia, 139 from government and 73 from intermediaries). These experts had the firsthand experience of working in the IUGPs in Qatar.

7.2.2 Conducting the Survey

The survey was hosted online for two months (April and May- 2018) on Survey-Monkey. The potential respondents were invited through emails and telephones and face-to-face meetings to take part in the study. The respondents were reminded twice to take part in the study before stopping the survey. All data was collected electronically.

The responses obtained from the interviews conducted earlier were also coded and integrated with the survey responses, and both survey and interview responses were analyzed together. Beside the 18 interviews, the total number of complete and useful responses received from the survey were 74. Hence, a total of 92 responses were considered in the analyses.[2]

---

[2] All respondents did not answer all questions; there were some questions which were added after the interviews and therefore the response from 18 interviewees for those questions is not available. Also, some of the questions were more specific than the others, e.g., there were two questions which could be answered by only those who had experience in both national and international IUGP settings. Hence, the respondents with experience in only national or international IUGP settings could not answer these questions. We will specify the number of respondents for those questions which have different number of responses than the total number of responses, i.e., 92. Hereafter, whenever the number of respondents have not been noted, it implies n = 92 (all respondents answered the question).

7.3   Results and Discussion

7.3.1   Demographics

The number of male respondents in the survey is quite higher than the number of female respondents (Fig. 7.1a); there are 74 male and 18 female respondents. However, this is not surprising since the country's mix of genders is similar to this proportion (approximately 75% males and 25% females) (World Population Review, 2019). Likewise, the number of nationals (usually referred as Qataris) who partici-

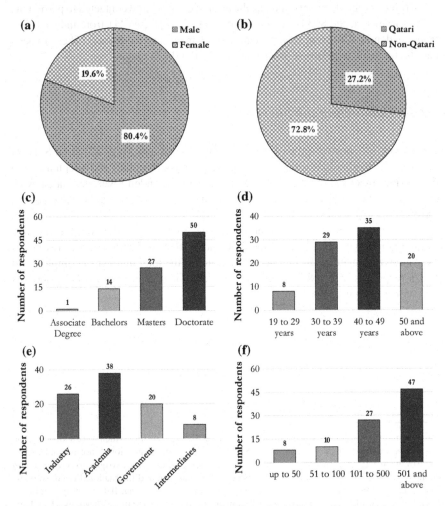

Fig. 7.1 Survey demographics; **a** gender, **b** nationality, **c** level of education, **d** age groups, **e** type of the sector, and **f** size of respondents' current organization (number of employees)

pated in the survey are significantly lower than the non-nationals (usually referred as non-Qataris); there are 25 nationals and 67 non-nationals (Fig. 7.1b). This is also somewhat representative of the mix of nationalities in Qatar, where nationals make up only around 12% of the total population[3] (World Population Review 2019). Among the non-nationals, most of the respondents are from North America, Europe, and Asia Pacific. Since Gremm et al. (2018) identified a big divide in the opportunities, approaches, and motivations of the nationals and non-nationals in Qatar, it is of significant interest to empirically view the difference of opinion of nationals (Qataris) and non-nationals (non-Qataris).

As far as the education of the respondents is concerned, there are 77 respondents in the survey with a graduate degree (Fig. 7.1c). More specifically, there are 50 Ph.D.s in the study, which shows a high number of responses from intellectually mature people, who at some stage in their career have gone through the difficulties of engaging with IUGP actors and should therefore be aware of the dynamics of these systems. The respondents' discipline of education ranges from social science to engineering, economics to sustainability, and medicine to law. Moreover, the age range of the respondents, shown in Fig. 7.1d, indicates the inclusion of more mature respondents in the sample.

Most of the respondents in the survey are from the academic sector ($i = 38$),[4] as shown in Fig. 7.1e, followed by industry ($i = 26$), government ($i = 20$), and intermediaries ($i = 8$). The low number of responses from intermediaries is because of two main reasons: (i) there are not many intermediaries in Qatar; and (ii) a few intermediaries which were contacted to take part in the study did not respond to the request.

Although we sought a higher number of responses from the SMEs (to understand their concerns and problems) but unfortunately, their numbers remain low (Fig. 7.1f). We believe that the primary reason for the low representation of small enterprises is simply the lack of small industries in the country, which was also identified as one of the major concerns during the interviews and will be discussed in detail later in this chapter.

Interestingly, the respondents in the 'industry sector' (Fig. 7.1e) are equally divided over the national and international organizations (Fig. 7.2); 13 respondents each from the national and international organizations. On the other hand, as expected, most of these respondents are from the oil, gas and energy sector ($i = 15$), which is the economic backbone of the country. Besides that, there are 4 respondents from the industrial sector and 2 respondents from the materials and chemical sector,

---

[3]Probably, if the blue-collar workers, who are mainly in the country for the FIFA 2022 World Cup-related projects only, are not considered in the mix of population (Gulf Research Center 2017), the number of nationals would reflect similar to what has been observed in the survey.

[4]Out of the total number of respondents who answered the question, i.e., 'n', 'i' represent the number of respondents choosing a particular choice, which in this case is the 'academic sector' among all other sectors. As mentioned before, 'n' will be noted only if it is different than the total number of responses, i.e., 92; where 'n' is not specified, n = 92.

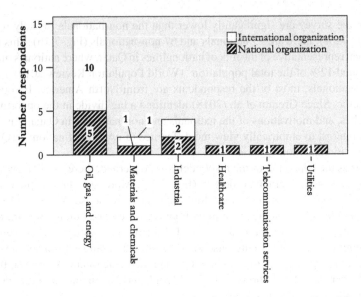

Fig. 7.2 Type and ownership of respondents' organizations in industry sector (n = 26)

followed by one respondent each from healthcare, telecommunication, and utilities.[5] Unfortunately, there are no respondents from the consumer discretionary, consumer staples, financial, information technology, and real estate sector. The reason why we do not have respondents from these sectors is because some of these sectors are virtually non-existent in the country while others have very limited presence.

Most of the respondents from the 'academic sector' (Fig. 7.1e) identified postgraduate studies and research as core activities of their institutions (i = 31), whereas only 5 respondents recognized the core instructional focus of their institutions (Fig. 7.3). Furthermore, there are two respondents who identified both, research and instruction, as core activity of their institutions. Academics from national universities (i = 15) seem to be more interested in participating in this study than the ones at IBCs (i = 6), but that is probably because of the lack of graduate degree programs offered at IBCs in Qatar which makes most of these campuses less interested in IUGPs. The involvement of research institutions and other centers is also appropriate in this study with 13 experts participating in the survey. In addition, there are 4 respondents from the colleges in Qatar who participated in this study. Since academic respondents are from a wide and diverse range of institutions, it will be interesting to analyze how they view the role, status and future of IUGPs in Qatar.

Similar to the respondents in the industry sector, the respondents in the 'government sector' (Fig. 7.1e) are also equally divided over the ministries (and the supreme councils) and semi-government agencies, such as QF (Fig. 7.4); 10 respondents each from the ministries (and the supreme councils) and semi-government agencies.

---

[5]For comprehensive definition of each industry classification, please visit https://www.msci.com/gics.

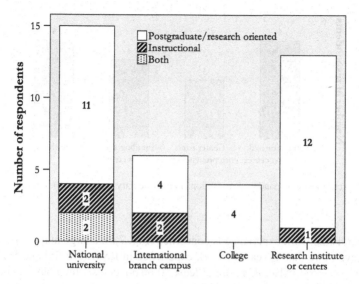

Fig. 7.3 Type and core activities of respondents' institutions in academic sector (n = 38)

Fig. 7.4 Type of respondents' institutions in government sector (n = 20)

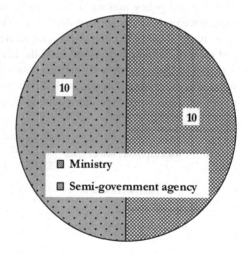

Considering that the contact information of government employees is generally not available on the websites in Qatar, and also that not all government employees have knowledge pertaining to the IUGPs, 20 respondents are adequate to represent the part of government interested in partnering with other actors of the knowledge economy.

The respondents from the intermediary sector (Fig. 7.1e) identified the involvement of their institutions in various activities (Fig. 7.5), including innovation and incubation (i = 5), entrepreneurship (i = 3), funding and support centers (i = 4). There are 5 responses for 'other' activities, which included dealing with free zones and facilitation in IP and commercialization activities.

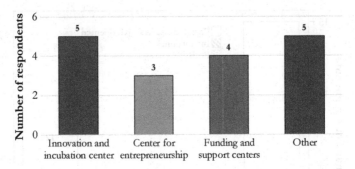

Fig. 7.5 Activities of respondents' institutions in intermediary sector (n = 8) (multiple responses possible)

As far as the work experience is concerned (Fig. 7.6), a large number of respondents have more than 5 years of work experience in R&D (i = 55) and leadership positions (i = 53). Also, there are at least 80 respondents which have some experience in the industry-academia collaborations. On the flip side, the respondents have least experience in entrepreneurial activities (i = 31 for no experience) and planning and policy making (i = 21 for no experience). The lack of respondents' experience in these categories is due to the following two reasons: (i) academia is not actively involved in the entrepreneurial activities and planning and policy making and most of our respondents are from the academic sector; and (ii) entrepreneurial activities in Qatar are virtually non-existent—there is a huge reliance on government for almost everything (Gremm et al. 2018). Most of the respondents participating in the survey have work experience outside Qatar and therefore are in a good position

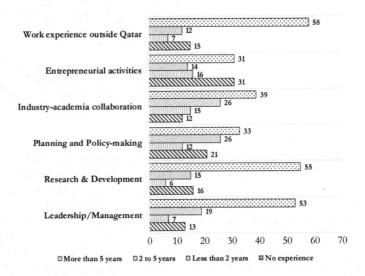

Fig. 7.6 Work experience of respondents

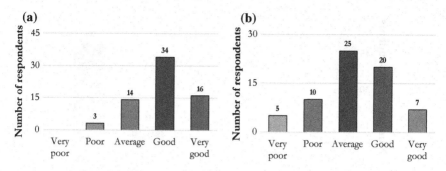

Fig. 7.7 IUGP understanding of the respondents (n = 67); **a** general understanding of IUGP mechanism, and **b** understanding of Qatar's IUGP mechanism

to compare the IUGP settings in Qatar with other parts of the world; 58 respondents have more than 5 years of work experience outside Qatar whereas 15 respondents did not work overseas.

7.3.2 Respondents' Understanding of IUGP

Respondents' general understanding about the IUGP mechanism is "good" (Fig. 7.7a); only 3 respondents choose "poor" understanding among 67 who responded to this question. In contrast, 15 respondents choose between 'poor' and 'very poor' when it comes to their understanding of the IUGP mechanism in Qatar (Fig. 7.7b). The number is higher than what was expected since, during the sample selection, we purposely contacted only those professionals who were likely linked to the IUGP settings in Qatar. This discrepancy highlights a strong need for promoting the awareness of IUGP mechanism in Qatar since even the professionals who deal with IUGPs on a frequent basis are not fully aware of the mechanism of these partnerships. Nevertheless, an overwhelming majority of the respondents acknowledged the importance of IUGPs for the scientific and technological progress of a country or a region (Fig. 7.8).

7.3.3 Quality and Preparedness of Qatar's IUGP System

Most of the respondents (i = 47) identified both national and international collaboration(s) of their organization with other IUGP actors in the past five year (Fig. 7.9); respondents from the academic sector seem more interested and active in forming the national and intentional collaborations (i = 22). At the same time, there are 32 respondents who identified only domestic collaboration(s) of their organization in the past five years. Interestingly, the number of nationals in the cross-nation partnerships

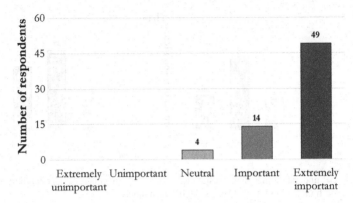

Fig. 7.8 Importance of IUGP for the scientific and technological progress of a country or a region (n = 67)

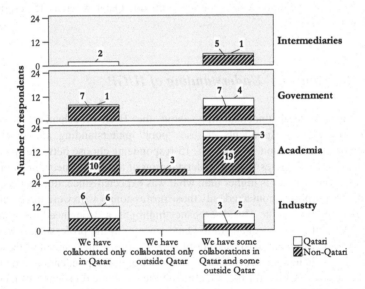

Fig. 7.9 Domestic and foreign IUGP collaboration(s) of respondents' organization in the past five years (n = 82)

(i = 13) are more than the nationals who collaborated with domestic institutions only (i = 9). Also, the 3 respondents who identified the collaboration of their organization with overseas IUGP actors only are all non-nationals in the academic sector. In terms of interpretation, this is a positive sign for the economic diversification of the country, as more locals are seeking global partners, and not for consultancy but to foster the domestic research infrastructure and to propel the drivers of knowledge ecosystem in the country (Thier 2017). For the respondents who identified both national and international IUGP partnerships of their organizations in Fig. 7.9, the extent of domestic and foreign partnership is about the same (Fig. 7.10); there are 21 respondents each for 'less than 50%' and 'more than 50%' of local collaborations.

However, when it comes to the satisfaction from the local partnerships the respondents who answered 'less than 50% of local collaboration' seem less satisfied in comparison to the respondents who choose 'more than 50% of local collaboration' (Table 7.2). For example, all 4 respondents who ranked local partnerships as highly unsatisfactory (in comparison to the international partnerships) belong to the 'less than 50% of local collaboration' group, whereas the only 2 respondents who ranked local partnerships as highly satisfactory (in comparison to the international partnerships) belong to the 'more than 50% of local collaboration' group. There can be two possible explanation for this trend: (i) there can be a bias associated with the extent of the local partnership, i.e., the respondents who believe that local capacity is sufficient to fulfil the domestic need probably form more local partnerships than the ones who believe that the local talent and resources are not sufficient; and (ii) the respondents in the 'less than 50% of local collaboration' group had more opportunity to interact with the international actors and hence could better compare their level of satisfaction from the local and international partnerships. It seems that the latter is true because the respondents, in general, identified a wide gap between Qatar's IUGP practices in comparison to the best practices in other countries where they have worked (Fig. 7.11); 32 respondents choose between 'poor' and 'extremely poor' when comparing Qatar's IUGP settings with other parts of the world. On the other hand, there were 7 respondents who choose 'better' and 'extremely better' for Qatar's IUGP settings. Admittedly, the difference of opinion was partly due to the subjective nature of the question; one of the interviewees mentioned that comparing Qatar to US would be different than comparing Qatar to other countries in the region, which is also in line with the results of Gremm et al. (2018). Nevertheless, the 3 respondents choosing 'extremely better' in this instance are all Qatari nationals.

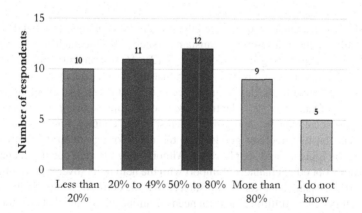

Fig. 7.10 Extent of local collaborations as a percentage of total collaborations—only for those respondents who identified both domestic and international collaboration of their organization in Fig. 7.9 (n = 47)

Table 7.2 Respondents' satisfaction from the local partnerships as a function of the extent of local collaborations (n = 42)[a]

Percentage of local col-laboration	Highly unsatisfied	Unsatisfied	Neutral	Satisfied	Highly satisfied	I do not know
Less than 20%	3	–	3	4	–	–
20–49%	1	1	5	3	–	1
50–80%	–	3	2	6	–	1
More than 50%	–	–	1	4	2	2

[a]The numbers in the table represent number of respondents, e.g., there are 3 respondents who are highly unsatisfied with the local partnerships which were less than 20% of their organizations' total partnerships (in comparison to their organizations' international partnerships which were more than 80% of the total collaborations)

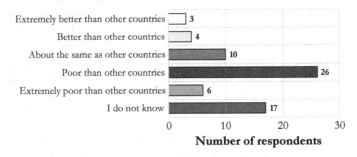

Fig. 7.11 Comparison of Qatar's IUGP practices with the best practices in other countries (n = 66)

Most of the respondents questioned the preparedness of country's IUGP ecosystem to deal with the unforeseen circumstances, such as the blockade imposed against Qatar by the neighboring countries (Fig. 7.12); 16 respondents, among 49 who answered this question, are of the opinion that the country needs to better prepare its IUGP ecosystem to tackle the unforeseen circumstances. The developing IUGP ecosystem in Qatar was also identified as an opportunity during the interviews to enhance the local capacity and global interest in the country. Since Qatar used to rely heavily on its import of goods from the neighboring countries, the blockade disrupted the logistics and supply chain. Although the gas-rich country managed to immediately opt other channels of import with the help of its sovereign wealth funds (Vohra 2019), usually, in these circumstances, it is the domestic setup which supports the overall system in fulfilling the local needs (Lamine 2005), and that is exactly why the IUGPs are of special importance to Qatar; these partnerships can help the local entrepreneurs in directly feeding to the needs of the country—be it food and water security, clean energy, or cyber security.

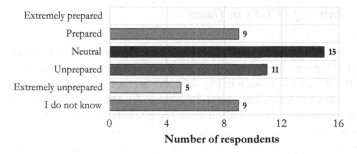

Fig. 7.12 Preparedness of IUGP eco-system in Qatar to tackle the unforeseen circumstances (n = 49)

In the as-is situation, the respondents believe that it is the government and academia who are playing the major role and have greater impact through IUGPs in comparison to the industries (Fig. 7.13); notwithstanding, the impact is somewhat medium even for government and academia. However, for the successful future of IUGPs in the country, the respondents see a greater role and impact of industry in these partnerships in comparison to that of the government and academia. A positive conclusion for academia, as depicted in Fig. 7.13, is that the respondents do not see much difference in the role and impact of academia in the current and future settings, which is a sign of satisfactory performance of academia in the current settings.

Fig. 7.13 The role and impact of industry, academia, and government in IUGPs in the current and future settings (n = 91 for current settings and n = 74 for future settings)

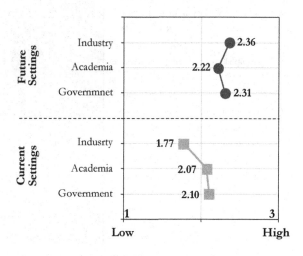

7.3.4 Status of IUGPs in Qatar

In order to further explore and understand where Qatar needs to improve in terms of its IUGP settings, the respondents were asked for their level of satisfaction[6] over various IUGP factors. As seen in Fig. 7.14, the average response of the respondents did not reach a full satisfaction level, i.e., 4 or 5, for any of the IUGP factors.

Nevertheless, the factors with more average satisfaction than 'neutral (=3)' are towards the positive side of the satisfaction spectrum while the factors with less average satisfaction than 'neutral (=3)' are towards the negative side of the satisfaction spectrum (Table 7.3). It can be observed that the factors towards the positive side of the satisfaction spectrum are somewhat related to the availability of resources, such as financial, human, and institutional, whereas the factors towards the negative side of the satisfaction spectrum partly represent the expectations from, and administration and management of, these partnerships, such as commercialization, outcomes, regulations and culture. This implies that while there are ample resources available for establishing and sustaining IUGPs in Qatar, the resource planning and management and the usefulness of the outcomes are not acceptable. The argument can further be explained by considering the two extremes of the satisfaction spectrum (Table 7.3): while the country ensures fairly high availability of funding for IUGPs (distance of

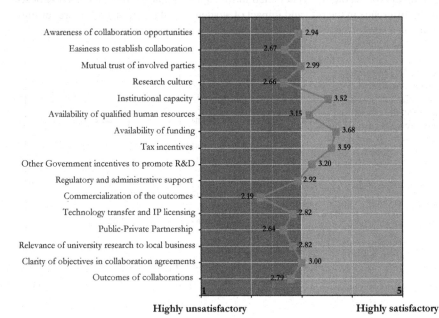

Fig. 7.14 Average satisfaction of respondents over various IUGP aspects in Qatar (for tax incentives n = 86 and for public-private partnership n = 74)

---

[6]1 = highly unsatisfactory; 2 = unsatisfactory; 3 = neutral; 4 = satisfactory; and 5 = highly satisfactory.

Table 7.3 Distance of the average satisfaction level of the respondents from the neutral satisfaction level for various IUGP factors (for tax incentives n = 86 and for public-private partnership n = 74)

Factor	Distance of average satisfaction level from 'neutral'	Factor	Distance of average satisfaction level from 'neutral'
Availability of funding	0.68	Regulatory and administrative support	−0.08
Tax incentives	0.59	Technology transfer and Intellectual property licensing	−0.18
Institutional capacity	0.52	Relevance of university research to local business	−0.18
Other Government incentives to promote R&D	0.2	Outcomes of collaborations	−0.21
Availability of qualified human resources	0.15	Easiness to establish collaboration	−0.33
Clarity of objectives in collaboration agreements	0	Research culture	−0.34
Mutual trust of involved parties	−0.01	Public-Private Partnership	−0.36
Awareness of collaboration opportunities	−0.06	Commercialization of the outcomes	−0.81

average satisfaction level from 'neutral' = 0.68), the commercialization of the outcomes from these partnerships is least satisfactory (distance of average satisfaction level from 'neutral' = −0.81). Similarly, while Qatar offers immense tax incentives for forming IUGPs (distance of average satisfaction level from 'neutral' = 0.59), which is basically due to the tax-free systems of the country (PwC 2016), the formation of, and outputs from, PPPs are unacceptable (distance of average satisfaction level from 'neutral' = −0.36).

While the average satisfaction from all aspects of IUGP in Qatar (Fig. 7.14) is about neutral (i.e., 2.97), the respondents identified various major initiatives that have fostered and/or supported the IUGPs in the country. Some of these identified major initiatives are tabulated in Table 7.4. While it does not qualify as an initiative, it is worth noting that various respondents identified the commitment of the leadership of Qatar as the primary support for the IUGPs in the country. On the other hand, some respondents recognized it as an individual effort and ambition to form these partnerships in absence of any external support. According to these respondents, there

Table 7.4 Major initiatives that have fostered and/or supported the IUGPs in Qatar

Academia (n = 26)	Industry (n = 22)	Government (n = 14)	Intermediaries (n = 5)
• Establishment of QNRF and NPRP funds • Establishment of the international branch campuses and national research centers (such as QEERI, QBRI, QCRI) • The mandatory involvement of industries in QNRF proposal and grants • The introduction of matching fund (by QNRF) in NPRP cycle 10 where QNRF pooled in the same amount of funds which the industries were willing to offer • Industry support for endowed chairs • The innovation lab started by the Ministry of Transport and Communications (MOTC) • Innovation internship in QSTP • Workshops and networking events organized by the industries, such as Qatar Rail	• Qatar National Vision 2030 • Establishment of Education City (and bringing the international branch campuses to Qatar) • Establishment of QF, QSTP, QNRF, and NPRP grants • Establishment of research excellence centers in Qatar University (e.g., gas processing center and advanced metallurgy center) • Some organizations have formal policy to collaborate with academia before going to consultants (where applicable and practically feasible) • Workshops, networking events, and career fair • Scholarships, sponsorships, and internships	• Qatar National Vision 2030 and Qatar National Development Strategy • Establishment of Qatar Foundation • Establishment of QNRF and NPRP funds • Establishment of QDB, QSTP and Qatar Chamber • Co-funding requirement in NPRP grants (cycle 10) • Giving NPRP funding priority to collaborative projects • Awarding NPRP to clusters • Qatar Food Security program	• Qatar National Research Strategy and Qatar National Manufacturing Strategy • Establishment of QF, QSTP, QNRF, and NPRP grants • Investment forums and cooperation networks of R&D centers and industries • Establishment of Manateq (free zones) • Research to startup program of QSTP • Product development funds offered by QSTP

are no major initiatives that are inclusive and bring together industry, government and academia over a single point research or innovation agenda.

Most of the respondents acknowledged the collaboration of their organization with other IUGP actors in the past five years (Fig. 7.15); 77 respondents, including all respondents from the intermediary sector, recognized partnership(s) of their firm with other IUGP actors. Most of the respondents who indicated no partnership in the past but potential partnerships in the future are from the government sector (i = 3), whereas, most respondents who choose no past collaboration or potential future

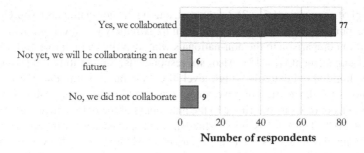

Fig. 7.15 Collaboration of respondents' organization with other IUGP actors in the past five years

collaboration with other IUGP actors are from the industry sector (i = 6). While the interest of government entities in forming IUGPs indicates potential of development of these partnerships in the future, at the same time, the deliberate separation of industry from these partnerships represents a big threat to the knowledge ecosystem in Qatar, as identified in the interviews as well.

The 77 respondents who acknowledged the collaboration between their organization and other entities in the past and the 6 respondents who indicated potential future collaborations, showed a medium-average degree of collaboration between their institute and other IUGP actors (Fig. 7.16). Interestingly, among the nationals (Qataris), more respondents are inclined to a medium to extremely high level of collaboration (i = 20) in comparison to the nationals who identified a low to extremely

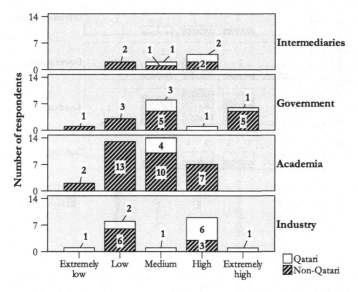

Fig. 7.16 Extent of collaboration between respondents' organization and other IUGP actors (n = 83)

low level of engagement (i = 3). In contrast, among the non-nationals (Non-Qataris), more respondents are inclined to a medium to extremely low level of collaboration (i = 43) in comparison to the non-nationals who identified a high to extremely high level of engagement (i = 17). Also, six respondents, representing the government entities, highlighted extremely high level of collaboration with other IUGP actors. This shows the disposition of government organizations towards establishing more partnerships in order to contribute to the attainment of the optimistic goals set out in the QNV 2030. This was also highlighted during the interviews with the government officials that the ministries and councils are increasingly looking forward in forming local partnerships which can support the development of partnering institutions on one hand and raise the knowledge quotient of the society on the other.

Besides the extent of collaborations, the satisfaction from the outcomes of these collaborations is also about average (Fig. 7.17); 29 respondents, out of 82 who answered this question, indicated a neutral satisfaction level whereas the respondents on the satisfactory and unsatisfactory side of the scale are nearly equal, i.e., i = 26 for unsatisfactory and extremely unsatisfactory and i = 27 for satisfactory and extremely satisfactory. Also, the responses from the nationals and non-nationals are not significantly different. Nevertheless, there are more extremely unsatisfactory responses (i = 5) than the extremely satisfactory responses (i = 2).

While the extent of collaborations and satisfaction from these collaborations are about average, the respondents showed overwhelming interest in forming such collaborations in the future (Fig. 7.18a); 59 respondents indicated willingness to form

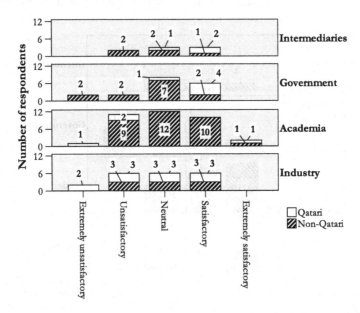

Fig. 7.17 Respondents' satisfaction from the outcomes of collaboration(s) between their organization and other IUGP actors (n = 82)

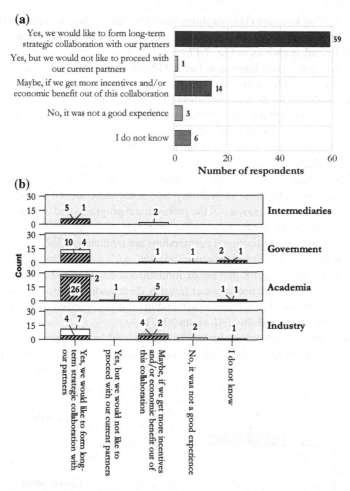

Fig. 7.18 Respondents' interest in forming similar collaboration(s) in the future (n = 83); **a** total responses, and **b** responses based on sector and gender

long-term strategic collaboration with their partners, which is a positive sign since formation of long-term partnerships is the only way to create impactful value-driven collaborations (Kothandaraman and Wilson 2001). On the other hand, there are 14 respondents who associated financial benefits with the future of these partnerships; these respondents showed an interest in establishing similar partnerships with other IUGP actors only if these collaborations result in economic benefits. Interestingly, among these 14 respondents 12 are male (Fig. 7.18b). This result is in line with the general body of literature where males' responsiveness to financial incentives is greater than that of the females' (Fryer et al. 2008). Similarly, the only respondent who showed interest in future collaborations but not with the current partners is a female from the academic sector. Nevertheless, there are 3 respondents who showed a red flag to future collaborations since their experience was not good. These 3 respondents are the ones who shared the extremely unsatisfactory view about the outcomes of collaborations in Fig. 7.17.

While we saw a keen interest in the IUGP actors to form more partnerships in the future (Fig. 7.18b), there are only 31 respondents who identified policies or incentives designed by their organization to promote partnerships with other IUGP actors (Fig. 7.19). Among these 31 respondents, most respondents are from the academic sector (i = 14) and the least respondents are from the industry sector (i = 3). All 3 respondents from the industry are Qatari nationals, which shows that either non-nationals are not aware of the organizational programs as much as the nationals or industries in Qatar do not offer many (internal) programs to promote such partnerships. The latter is probably true since there are 10 nationals (in the industry sector) who are unaware of the promotional programs and incentives offered by their organizations. Some sector-wise examples of the policies, programs, and incentives to promote institutional partnerships are tabulated in Table 7.5. While all sectors need to device more programs to promote such partnerships in the future, it is clearly the industry sector that needs fundamental changes and improvements in the organizational culture and mindset towards these partnerships; it is not possible to achieve excellence through these partnerships if it is perceived in the industry as corporate social responsibility (Al-Mana 2017).

Furthermore, the respondents have put a huge weight on the importance of improving the research culture in the country. As highlighted in Fig. 7.20, poor research

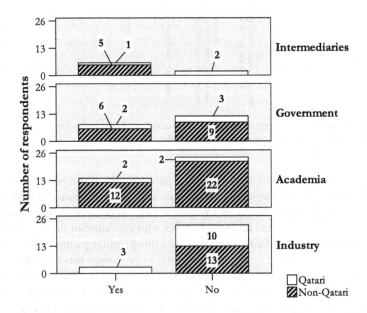

Fig. 7.19 Specific policies or incentives designed by respondents' organization to promote partnerships with other IUGP actors

Table 7.5 Policies, programs, and incentives designed by each sector to promote IUGPs in Qatar

Academia (n = 14)	Government (n = 8)	Intermediaries (n = 6)	Industry (n = 3)
• Faculty evaluation and profiling based on collaborations • Internal seed funding and matching NPRP grants (10th cycle only) • Endowed chairs and professorship program • Technical service facilities (allow government institutions and industries to have tests in university labs) • Building research portfolios • High impact research grant • Outreach and research office • Arranging networking events with industry and government partners	• PPP department and law • Written departmental objectives and strategies of ministries to collaborate with universities • Joint international funding (in the past) • Direct funding of innovative collaborative projects • Domestic co-funding (in cash or in kind) • Participation of research end-users in government-run collaborations • All NPRP projects are taken-up by the IP office and QSTP to seek opportunities for commercialization • Holding events and seminars for awareness and networking	• Product development funding • Industry support grant • Research to startup program (commercialization programs)	• Sponsorships for graduate students • Internships for undergraduate students • Corporate social responsibility

culture ($i = 57$) is by far the major challenge in the improvement of IUGP settings in Qatar, followed by the lack of awareness of collaboration opportunities ($i = 53$) and the difficulties in establishing the collaborations ($i = 42$), which is related to the bureaucratic hurdles in formalizing partnerships. Other issues include the lack of commercialization of the outcomes ($i = 36$) and the lack of regulatory and administrative support ($i = 36$) to deal with these issues. These are the same issues which were also highlighted in the previous chapter (Chap. 6). The lack of commercialization of the outcomes also poses question on the effectiveness of the operations of intermediaries in the country since it is particularly their task to take the research to an applied level. Another important factor, which was frequently emphasized during the interviews, is the lack of mutual trust of the involved parties ($i = 34$). In absence of mutual trust, it is not possible for the partnering institutions to get any benefit out

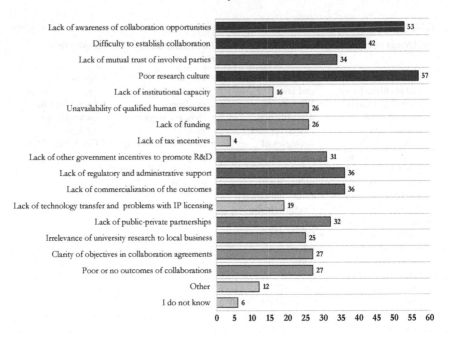

Fig. 7.20 Major issues and barriers in the improvement of IUGP settings in Qatar (n = 74 for lack of public-private partnerships) (multiple responses possible)

of these partnerships, rather in these circumstances the focus of partnerships shifts to 'who can dominate the collaboration', which then affects the interdependence and ultimately the commitment of organizations (Tuten and Urban 2001).

The lack of PPPs has also been identified as a major concern along with the lack of government incentives to promote R&D. These factors are extremely important from Qatar's perspective since most of the enterprises in the country are either owned or funded/supported by the government. Providing incentives can play a motivational role in founding effective IUGPs across the country. On the other hand, the private sector is limited to the oil and gas companies who are in this region for their business, and not research, innovation and knowledge creation, according to one of the interviewees. Also, what we have seen from the case studies of US, Singapore, and Norway (in Chaps. 2, 3 and 4 respectively), it is the PPPs and government incentives that form the bases of the private sector, knowledge-intensive businesses, and SMEs, which eventually create more economic opportunities on one hand and help in diversifying the national economy on the other.

The concerns of moderate extent to the respondents include: poor or no outcomes of collaborations (i = 27), lack of clarity of objectives (i = 27), lack of qualified human resources (i = 26), lack of funding (i = 26), and irrelevance of university research to local businesses (i = 25). Most of these issues have a local significance, e.g., the lack of qualified human resources is because of the low participation of nationals in the knowledge-based activities (Gremm et al. 2018), especially the males, whereas the non-nationals have less stake in the society, which is why their turnover is high. Also, the identified irrelevance of university research to local businesses is because universities are focused more on the oil and gas research, which is although related to the core economic activity of the country but this results in overlooking other local businesses (non-hydrocarbon), such as tourism, leisure, real estate, food and clothing, logistics, media and broadcast, airline and automotive, telecommunications, and manufacturing (McSparren et al. 2017; Ministry of Economy and Commerce 2016; Oxford Business Group 2015).

On the other hand, the concerns of relatively low importance to the respondents are tax incentives ($i = 4$), lack of institutional capacity ($i = 16$), and lack of technology transfer and problems with IP licensing ($i = 19$). Nevertheless, the institutional capacity, technology transfer, and IP licensing are associated with other factors which are discussed earlier, such as the culture, relevance, and outcomes of research, and regulatory and administrative support.

Some 'other' major hurdles in the improvement of IUGP settings in Qatar ($i = 12$) are breach of the data confidentiality, lack of the ability to deliver (the deliverables) within agreed timeframe, lack of motivation and commitment beyond the financial gains, lack of patience, expectation of immediate (financial) returns, lack of governmental regulations/instruments to enforce protected time, lack of instruments to prevent misdirection of funds, institutional politics, lack of contribution of intermediaries, lack of communication between and within organizations, funding of less attractive projects (with high capital and operational expenditure), and language and cultural barriers.

7.3.5 What Does Non-academic Actors Think About the Role and Impact of Academia in Qatar's Knowledge-Based Economy

The three non-academic actors of knowledge ecosystem in Qatar, i.e., industries, government, and intermediaries, find clear relevance of collaborating with academia for the business of their organization (Fig. 7.21); most of the respondents in the government sector ($i = 10$) identified the collaboration with academia 'extremely relevant' and most of the industry representatives ($i = 10$) marked these collaboration as 'relevant'. This is an approval of the quality of the academic work from the non-academic actors, which highlights the prospects of academic excellence in the country. Nevertheless, there are some significant improvement opportunities for the academic actors in Qatar which will be discussed later in this chapter.

Among the (non-academic) nationals, more respondents think about academic collaborations as 'neutral' to 'extremely relevant' for the business of their organizations ($i = 19$) in comparison to the nationals who identified it as irrelevant to extremely irrelevant ($i = 0$) (Fig. 7.21). In addition, all 5 respondents who marked academic collaborations as 'extremely irrelevant' are (non-academic) non-nationals.

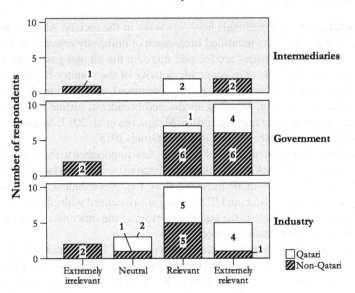

Fig. 7.21 Relevance of collaborating with academia for the business of non-academic respondents' organizations (n = 44)

While most of the respondents in industry and government sector believe that collaborating with academia is relevant to the business of their organization (Fig. 7.21), academia is not the first choice for these organizations when it comes to acquiring high-tech products, processes, services, and technologies (Fig. 7.22); 25 respondents prefer to procure high-tech solutions from overseas and 9 respondents favor the in-house development. The most common reason for not choosing academia for the development of high-tech solutions is the time constraint (i = 23), as shown in Fig. 7.23. Issues pertaining to time constraints were also highlighted during the interviews; one of the interviewees mentioned that his organization can procure the required high-tech product or service in the same amount of time which takes to

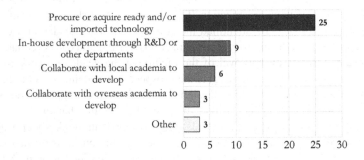

Fig. 7.22 Preference of industry and government actors for acquiring high-tech product, process, service, and technology (n = 47)

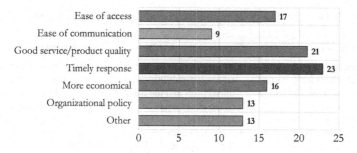

Fig. 7.23 Reason of not preferring academia for acquiring high-tech product, process, service, and technology (n = 38) (multiple responses possible)

merely establish a partnership with the academic institutions, which is why procuring is preferred over partnering. Other major reasons for bypassing local academia for high-tech solutions are quality (i = 21), ease of access (i = 17), and cost (i = 16).

Some of the major problems which the respondents in the three non-academic sectors experienced in their organizations' past collaborations with academia are tabulated in Table 7.6. From the comments it seems that the non-academic actors have lost their trust in academia primarily due to miscommunications and mis-commitments. The high turnover rate in academia has also been a major setback to long term commitments. Another complain that was frequently and frankly mentioned during the interviews is the overemphasis of academic institutions in publishing articles in academic journal rather than diversifying their research streams to solve the local grand challenges. This calls for a common understanding and probably a mutual change in that publishing is a requirement in academics; if academics do not publish it would be counterproductive for their career as faculty. Academia in Qatar may accept academic work which are industrially relevant and commercially oriented as part of their career development in return for industry sector to allocate a certain percentage of funding for academic research relevant to their business. Naturally, such a mutual understanding and agreement can only be realized and sustained with government participation through regulatory facilitation, incentives, and recognition.

7.3.6   What Does Academia Think About the Role and Impact of Non-academia Actors in Qatar's Knowledge-Based Economy

First of all, academics showed confidence in the relevance of the core activities of their institutions, whether research or teaching or both, to meet the needs of the non-academic actors in Qatar (Fig. 7.24); 21 respondents identified the core activities of their institutions as 'relevant' and 9 respondents identified these as 'extremely relevant'. Nevertheless, two respondents identified the core activities of their aca-

Table 7.6 Problems faced by the non-academic actors in collaborating with academia in the past five years

Industry (n = 15)	Government (n = 12)	Intermediaries (n = 5)
• Administrative processing in academia is long, cost of research is high, and results are not quick • Industry is used only for funding (some believe that industries are exploited to only get the QNRF grants) • Academic research does not create any value for businesses • Untimely delivery of outcomes and lack of commitment are two major academic problems • Confidentiality and rights of ownership of IP are big issues • Focus of academia is publishing in journals rather than solving real problems • Sometimes the projects/collaborations end without any deliverables • Besides engineering, academic research (in finance, supply chain and other fields) is virtually nonextant • The academic politics makes it difficult for external partners to collaborate • Academics are unaware of the local industry needs and national challenges • Flow of information and data is not smooth • There is frequent change of personnel which creates doubts over long-term commitments • Poor coordination, misunderstandings, and difference in priorities • Academia is not trustworthy	• Mismatch between the research capabilities of the universities and the needs of the government institutions • Academia does not address the local grand challenges, rather they are more interested in international problems • Government partnerships with academia lacked industry participation due to the distrust of industry on academia • Academic research is more fundamental than applied • There is a lack of research culture in academia • The academic priorities change very frequently • Academia is not fully aware of the collaboration opportunities • Education system in Qatar does not support research • Participation of nationals in academic research is extremely low • Academia needs to understand the budgetary constraints and must tailor their research activities to income generating outcomes • Limited availability of diversified faculty and staff (too much focus on engineering) • Poor communication and uncertainty due to frequent change in personnel • Lack of follow-up (sometimes projects stop midway without any notification or outcome)	• High turnover rate makes academic research non-resilient, i.e., if one person leaves the institute, the whole project collapse • The focus of academia is publishing and not the commercialization of the research outcomes • The quality of academic research outcomes is not sellable and that is why there have not been noticeable spin-offs • Lack of a formal process for the engagement of intermediaries with academics over the course of research projects • At an individual level, there is a lack of incentive, and hence motivation, to apply research outcomes to solve practical issues • Benefits of collaborations are not clear • Lack of legal framework for IP rights • Lack of infrastructure for clinical trials • The collaboration opportunities are not advertised properly

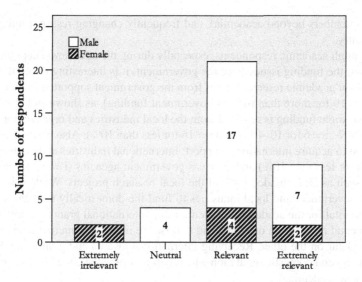

Fig. 7.24 Relevance of the core activities of academic institution for the business of non-academic organizations (n = 36)

demic institution as 'extremely irrelevant' for other IUGP actors; both respondents are females.

The academic respondents showed an unsatisfactory opinion about the quality and quantity of IP created as a result of IUGPs in the past five years (Fig. 7.25); only 3 respondents are satisfied with the IPs resulting from the partnerships whereas 10 respondents showed unsatisfactory and highly unsatisfactory results. However, during the interviews, it was highlighted that the lack of IP creation is associated with low funding, lack of motivation of other partners in the partnerships, lack of incentives for the faculty (such as sabbatical leaves and economic opportunities for

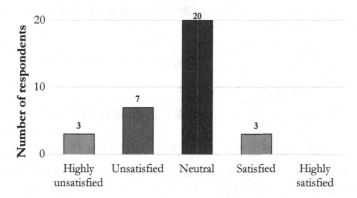

Fig. 7.25 Satisfaction of academics with regards to the quality and quantity of IP created as a result of IUGPs in the past five years (n = 33)

faculty members beyond academia), and frequently changing research interests of the country.

Although academic respondents, especially during the interviews, kept mentioning about the funding issues from the government, it is interesting to see that most funding for academic research comes from the government supported funding agencies (i = 19 for more than 61% of government funding), as shown in Fig. 7.26. In addition, some funding is reported from the local industries and organization (i = 4 for 41–60%, i = 6 for 10–40%, and i = 16 for less than 10%). Also, a few academics managed to acquire international support; international industries and organizations (i = 16 for less than 10%) and overseas government agencies (i = 12 for less than 10%), such as NSF, funded parts of the local research projects. While it is critical for the government and local industries to fund the domestically relevant projects, it is also vital for the academics to secure more international grants to support their fundamental research on one hand and to raise the interest of international community in Qatar on the other. Receiving international awards and grants may help in building a good domestic reputation which will reinforce the confidence of the local partners in academia.

The problems which academic respondents experienced in their institutions' past collaborations with industry and government are tabulated in Table 7.7. Most of the respondents believe that it is the lack of trust and interest which holds back the partnerships between academia and industry from being successful. From the government's side, the academics expect flexibility in regulations and support in procuring research equipment and materials.

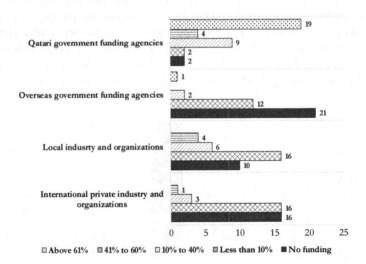

Fig. 7.26 Breakup of academic funding sources as a percentage of total funding secured by the institution in the past five years (n = 33)

Table 7.7 Problems faced by the academic actors (n = 30) in collaborating with industry and government in the past five years

Problems with Industry	Problems with Government
• Casual attitude towards the significance and role of research/research culture in the development of societies • Industrial partners act as consumers rather than partners in research and have unrealistic expectations from research outcomes • Local industries do not have R&D program because they do not trust in R&D and do not see a need for participating in R&D activities • Lack of planning, poor communication, lack of understanding of mutual expectations, and frequent change of management personnel • Industry needs quick solutions which is not possible through R&D—it requires time • Getting started' is the biggest challenge in working with industry because industrial actors are not aware of their own needs • There is a lack of diversity in the industry in Qatar which limits academics to partner only with oil/gas firms. Over and above, the national oil/gas companies are not interested in partnering with academia. • Industry actors do not trust on others, especially academics • Industrial support is sometimes cosmetic—supporting redundant seminars, talks, and workshops with no implementable outcomes. • The personnel who communicate with academia over forming and executing the partnerships do not have any authority, whereas the ones who have the authority do not communicate. This results in tremendous waste of time • Industry partners are always reluctant to share reliable data • The commercialization of research outcomes does not have to be directly beneficial for industry, rather it may be benefitting the larger community, municipality, and the nation • Amount of funding per researcher is not enough to make industrial breakthrough	• Policies and regulations change frequently, whereas research is a long-term activity • The current IP policies are counter-productive • Government entities do not actively engage in research, rather treats academics as consultants • Short term commitment and planning of government institutions (compared to US where policies stay for at least 15 years) • Lack of planning and delay in the release of funding • Faculty cannot be involved in commercialization of outcomes of research since the system do not allow them to leave university for 2–3 years (sabbatical leaves) • Poor management of protocols in unforeseen circumstances • Bureaucracy in government agencies and a lack of flexibility • Red-tapping, extensive paperwork, and unnecessary regulations, all result in waste of time • Procurement of the research equipment is not facilitated—"we *cannot be competitive if it takes two years for an equipment to arrive in our labs*" • Government organizations have a mindset of "build it and leave it" which fails the projects eventually due to a lack of continuous regulatory support

7.3.7   *Government and Intermediary Support*

The respondents from the government sector are more inclined towards the satisfactory side when it comes to their perception about the support of government institutions towards crucial elements of IUGPs, as shown in Fig. 7.27; there are more factors with higher average satisfaction level than the 'neutral (=3)', which shows a positive trend over the satisfaction spectrum. At the same time, it is quite possible that respondents from government sector might have taken a protective and reactive stand in their answers as they might have felt that they were responsible for IUGP laws and regulations. The factor with the highest average satisfaction level (i.e., 3.70) is R&D. Some respondents highlighted during the interviews that while government institutions support R&D in the country, other partners, especially academia, do not value these efforts and take it as an opportunity for increasing their research spending. On the other hand, the only factor which shows lower average satisfaction than 'neutral (=3)' is the commercialization of inventions. A mixed opinion was observed about this factor during the interviews; some respondents believed that it is not the task of government to commercialize the inventions, whereas others believed that government must form new institutions that not only monitor the spending of research funds but also the quality of the outcomes and potential of commercialization of these outcomes. The latter also believed that transparent research spending, good quality of research outcomes, and higher potential of commercialization of inventions must be linked with future grants and awards of the faculty and the institution.

The responses from the respondents in the intermediary sector are also towards the satisfactory side of the satisfaction spectrum when it comes to their perception about the support of intermediaries for various elements of IUGPs in Qatar, as shown in Fig. 7.28. Similar to what was observed through the average satisfaction of respondents (in all sectors) over various IUGP aspects in Qatar (Fig. 7.14), the respondents in the intermediary sector showed lower average satisfaction level than the 'neutral

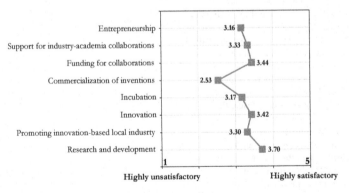

Fig. 7.27 Average satisfaction of the respondents from government sector about the support of government institutions towards selected elements of IUGPs (n = 16 for incubation and n = 20 for the rest)

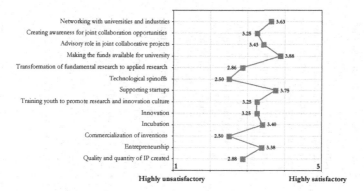

Fig. 7.28 Average satisfaction of the respondents from intermediaries about the support of intermediaries towards selected elements of IUGPs (n = 8)

(=3)' for the outputs (such as technological spinoffs and commercialization of inventions) and a higher average satisfaction level than the neutral for the inputs (such as making the funds available and networking with IUGP actors).

The respondents identified two important elements which are, although, beyond the control of intermediaries but have a significant effect on the success of their operations: (i) the good quality of the research outcomes; and (ii) access to a large end-user market. According to these respondents, the quality of research outcome is a significant input for the commercialization activity, and therefore, if the input is not up to the mark, the commercial product or service will also not be attractive. Similarly, beside the market of Qatar being small, the needs of this market are different from the research being carried out in the academic institutions, which eventually makes it difficult for intermediaries to launch a domestically relevant product or service. This shows a need for the enhancement of communication between academia and intermediaries on one hand and a more focused awareness and inclusion campaign from the intermediaries to attract more academics at an early stage of the research projects. Also, there is a strong need for an entrepreneurial intervention which has interest in commercializing the research outcomes.

7.3.8 Suggestions for IUGP Actors

All respondents made suggestions for improving the IUGP settings in Qatar. Collectively, these suggestions represent a holistic framework that the country should implement in the relevant facets. Table 7.8 gathers and summarizes a list of suggestions as received from all stakeholders to establish well-functioning, effective, and tailored IUGP strategies for Qatar.

Table 7.8 Suggestions for the IUGP actors in Qatar

Industry	Academia	Government	Intermediaries
• Industry should seek more information about the existing capacities in the universities and R&D centers • Increase research funding, especially for the non-hydrocarbon research, and provide more scholarships at the graduate level • Promote the use of university labs among local industries • Enhance communication with other actors through conferences and workshops • Provide incentives at department level and mid-management level to collaborate with academia • Develop a database of 'industry needs' and make it available to academics (open source) • Provide industry needs to research and IPTT offices • Industries should get involved in the awarded NPRP grants (they have to be equal or partial contributors and not merely the consumers)	• Outreach to industry must be improved through conferences and workshops • Arrange more visits to industry to share their knowledge, experience, and aspirations (relevant to Qatar) • Academia must maintain a database to showcase their capabilities • Undertake research to address local issues with applied outcomes • Increase public outreach (to explain the benefits of research outcomes) • Academics must take 'timely delivery of outcomes' more seriously • The motivation to form partnerships should be more than financial award • Academics must work on improving the research culture • Academics must find ways to deal with the internal politics of their institutions • Academics must raise the bar of research quality • Academia must absorb more nationals in the knowledge-based roles	• Arrange more interactions between industries and academia (increase awareness) • Encourage applied research by offering more funding • The collaborations between industries and academia and the research progress must be monitored by government entities (from establishing clear goals to timely delivery of deliverables) • New funds should only be released if previously awarded funds are justified • National research strategy should be revised in the light of the blockade, and a mechanism should be devised for future revisions in similar circumstances • Government should support clusters through incentives • Incentives (tax) should also be given to industry for collaborating with academia • Encourage nationals in the knowledge-based jobs • Give more rights to the non-nationals in order for them to realize a stake in the society • Reward technologists who provide new technologies	• Anticipate research outcomes in the industry and academia partnerships • Intermediaries must function as 'quality control department' for research outcomes • Create a market for private research and development consultancies • Advertise more frequently about their programs and capabilities (increase their visibility) • Creating startups should be made easier (allow university-based researchers to lead startup companies without quitting their jobs at the university) • Establish working groups to elaborate on their role • Interact with PhD students • These are the most critical entities in IUGPs and hence should better define their business model • Take the lead in creating collaboration schemes

(continued)

Table 7.8 (continued)

Industry	Academia	Government	Intermediaries
• Always agree on mutually beneficial goals before the start of the project • Provide support in commercialization • Hire qualified academics, especially PhDs • Increase collaboration with ministries and councils • Value research, increase trust and confidence in local capabilities • Must engage more with the general public regarding the industrial outcomes of research (through local media outlets and social media) • Industrial University' must be formed to train the nationals for specific needs of industries • Open door policy' to work with academics and government institutions (without the fear of sharing the data) • Industry must share the real business cases with academics to improve the classroom experience • Industry actors must think long-term and avoid looking for immediate benefits	• Academics must diversify their research interests (beyond engineering) • Academia must increase transparency in research spending • Research outcomes should create value to the society • Form more international collaborations • Increase the graduate programs in the country • Establish internal office of innovation and get the faculty involved in startups • Professionals from industry and government should be on the board of universities • Attract more qualified foreign talent • Universities must link the promotion of faculty members with the relevance of their research to the local market • Develop long-term relationship with industry • Universities should offer seed-funding for industry collaborations • Academia should form clusters to collaborate and address the local challenges (vicinity to education city is ideal for that)	• Impose less regulations on the import of research equipment and material (less red tapping) • Maintain a database with the list of opportunities/challenges that require IUGPs (such as opportunities in World Cup, Qatar Rail, and road accidents) • Communicate the success stories of IUGPs with general public • Government offices must train industry and academia for the upcoming PPP law • Involve academics and industrial actors in advisory roles, policy making, and setting up national R&D agenda • Enhance ease of doing business (less bureaucracy and paperwork) • Attitude needs to be changed—IUGPs are formed - not purchased, and therefore, get more local support • Government must bring the R&D personnel (living in the country for a long time) together and create a stake for these qualified people in the research ecosystem • Offer competitive (innovation) awards for graduate and undergraduate students	• Provide additional seed funding for tech transfer and entrepreneurial activities • Increase substantive collaboration with international counterparts • Initiate more forums to facilitate dialogues, discussions and negotiations between the IUGP actors • Offer tech-development internships

(continued)

Table 7.8 (continued)

Industry	Academia	Government	Intermediaries
• Involve academics in advisory roles (for developing strategies) • Industries must form their own research centers to collaborate with academia • Replace the foreign-based consultant culture with local collaborations • Promote higher education among the staff • Find easy ways to share reliable data • Relax the IP issues and be more openminded • Understand the limitations of academia (students, high turn-over, space and funds) • Invite and/or accommodate academics (faculty, scientists, researchers, and students) for short-long term work in industry	• For the IBCs, the faculty should get joint appointment at their Main Campus • Establish tailored continuing education programs for industry actors • Share knowledge, results, current activities, and capabilities through periodic workshops and meetings • Introduce collaborative training for undergraduate students • Bring (innovative) educational reforms—align curricula (at graduate level) with the needs of the industries in Qatar • Establish industrial advisory board (e.g., universities should plan graduate degrees in partnership with industries)	• Device a biding mechanism for awarding consultancy projects (between academic institutions and consultancy firms) • Provide a collaborative research scheme (outside NPRP) for specific industry projects • Provide funding for startup companies—even to the non-national entrepreneurs • Each government institute must allocate separate research funds to initiate collaboration with academia (to address their specific needs) • Device legal frameworks to protect SMEs • Further support (and ease) venture capitals and FDIs • Legislate to formalize the framework for IUGPs	

(continued)

Table 7.8 (continued)

Industry	Academia	Government	Intermediaries
• Cluster around certain common applied research needs with other companies • Offer more industry professorship chairs in the academic institutions • Industry must collaborate with academia beyond financial support • Local businesses (such as shopping malls, hotels, and sporting and recreation centers) must also involve in forming partnerships for added value • The number of private firms need to increase in the country, especially in product development sector (having a knowledge economy requires a very strong industrial economy)	• Be less rigid in rules of engagement—'open door policy' to work with industry and government • Universities must increase inter-university and inter-department collaborations • Academics should work beyond publications and do more for the application of research outcomes • Re-evaluate existing Memorandums of Understanding (MOU) and reconsider if these are needed • Universities must establish internal centers for innovation and entrepreneurship • Avoid duplication of research	• Develop research capabilities within government bodies • Government organizations should also seek global collaboration in order to boost the research culture • Provide more incentives to the academics (not only monetary but offer paid leaves to allow them to be involved in knowledge-intensive businesses) • Government bodies must re-evaluate their mechanism to evaluate project funding • Government funded research should be more strategic and should be evaluated based on the needs of the country • Government should increase research funding for competitive proposals	

7.4 Conclusions and Recommendations

Role and positive impact of industry-university-government partnerships (IUGPs) in innovation capacity building, knowledge ecosystem and economic development have been increasingly realized by countries across the world. As a result, all developed and developing countries make efforts to strengthen long-term, healthy and functional linkages between industry-university-government to create an environment that supports innovation, technology upgrade, and development of knowledge ecosystem. In the previous chapter, we performed a detailed study on the IUGP trends, motivations and mechanisms in Qatar and compared its framework and evolutionary pattern to the three advanced and emerging economies, including United States, Norway and Singapore. The comparison was based on the evolution of four key drivers of IUGP, which include: (1) institutional and cultural setting; (2) legislations and regulations; (3) intermediary structures; and (4) the public promotion programs.

Among the four economies, US has an advanced and mixed economy backed by abundant natural resources and huge FDIs. Singapore's economy is also largely powered through FDIs and infusion of international talents through carefully crafted immigration policies and incentives, whereas Norway and Qatar rely heavily on their natural resources while the latter mainly depends on low-semi-skilled international workforce. Despite the differences in social, cultural, political, and economic architecture of the four nations, the timeline of changes in the institutional structures and legislations and the development of new structures and promotional programs indicate a similar trend. The development follows a path that starts with strengthening of the absorptive capacity of the local firms followed by the technological upgrade of the industries (small to large and private to public) with applied research and high-tech innovation which can fulfil the needs of the society.

US has a very long and strong history of IUGPs, which helped the country in forming the current state of its advanced innovation system and knowledge-based economy. Both the state and federal government actively supported university-industry collaborations since early 1950s. By 1970, US already had a dedicated IUGP program, i.e., IUCRC. Currently, US has a strong network of advanced industries supported with a network of robust R&D physical and social infrastructure. Universities in US pioneered the research, innovation, and commercialization activities and are functioning equally good in both research and education. Industry-University linkages in Norway and Singapore have been initiated as a consequence of governments' general economic policies. Government support and prioritization of IUGPs accelerated in these countries the past two decades. The initial development of absorptive capacity among the local firms in both these countries relied on knowledge transferred through MNCs. Singapore, in particular, is performing exceptionally well in terms of its education quality and achievements; in a short span of time, national universities have gained international recognition. Besides, the country has made a huge investment in R&D infrastructure and is currently pacing the development of high-tech industries. The government policies and programs related to IUGPs in all three countries have been especially focused on entrepreneurship and R&D. The governments used various incentives to promote these partnerships across the countries, which include tax break, IP rights, economic incentives, and market support.

Qatar, on the other hand, appears to be in its early stage of knowledge-intensive development, and its IUGP settings predominantly have a static configuration. In general, there is a consensus among the experts on the need of an effective IUGP system in Qatar, but existing formulation, level, functionality, and impact of such partnerships is unsatisfactory. Therefore, industry and government opt for ready-to-purchase and ready-to-implement solutions, products, and technologies mostly from international sources, whereas these solutions can be developed domestically with a close and targeted collaboration with the local academic institutions, whose capacity is considered sufficient for the market size and needs of the country.

Although Qatar has made significant efforts and investments for diversified sustainable economic development, the effective utilization and implementation of these efforts are lagging. Presently, the partnerships between research entities and industries, and the participation of latter in these partnerships, are limited and weak. Industry and government's support for universities is somewhat limited to 'social responsibility', rather than mutually beneficial long-term research-oriented partnerships. Moreover, the main player within the industrial sector in Qatar is the oil and gas industry. Within the oil and gas industry, Qatar Petroleum (QP) makes the call for economic, business, financial, operational, and developmental activities. At the same time, QP does not have an R&D arm, which further limits the chances of domestic R&D collaborations in the entire oil/gas industry. Government, on the other hand, has not been proactive until recently when it brought major industrial players and government officials together under the umbrella of Qatar Research Development Innovation (QRDI) Council in 2019. The council is dedicated to enhancing Qatar's resilience and prosperity through locally empowered and globally connected research, development and innovation.[7]

Based on the case studies carried out in the first few chapters, results of the quantitative comparison of GII indicators of Qatar with other three countries, and the interviews and survey carried out with the IUGP actors in Qatar, we propose the following recommendations (in addition to the specific suggestions listed in Table 7.8):

- First and foremost, government of Qatar should lay the foundation for human capacity development from immigration, residency and citizenship perspective, allowing qualified professionals to obtain permanent residence, or even citizenship, through a transparent, progressive and selective immigration system. The naturalization will develop an interest of the foreign knowledge-intensive workforce to stay in the country for long-term. This will also alleviate a significant number of concerns raised during the survey regarding the high turnover rate, which is indeed a critical and long-term issue that affects the economic, social, cultural and financial efficiency of organizations and the entire country. At the same time, a well targeted immigration plan will increase Qatar's human capacity instantly by attracting more skilled and qualified people, who will support the

---

[7]http://qrdi.org.qa/.

country's vision for economic diversification by initiating and getting involved in various economic activities. Naturalization can be an extremely low-cost and low-risk solution, if managed properly, for increasing the domestic human resource capacity of Qatar.

- From a policy perspective, probably through the newly established QRDI, government should design and issue legislations to require all industrial and business entities in Qatar (above a certain threshold of size and revenue) to establish active R&D arms, starting from the national oil and gas company, QP. Firms should also be required to allocate a certain percentage of their revenues for R&D, either in-house or in partnership with local universities. In fact, a similar decree should be considered for some of the ministries and public agencies to allocate a certain percentage of their budget for R&D and small business innovation, either in-house or through university collaborations. Additional details of such decree can be designed around targeting, implementing, and monitoring how such budget/revenue allocations are utilized. Such a policy will establish a very strong fundamental basis for the rest of the recommendations.
- Government should design legislations to further require, facilitate and incentivize business organizations and industries to support and collaborate with academia and educational institutions by allocating matching funds, tax exemptions, land and other resource allocation, as well as IP and commercialization flexibilities. The legislation should make the participation of individuals (from industries and universities) easy in such partnerships, without putting their career to a disadvantage. This will promote productive and effective exchange of human resources between and among the IUGP players, and hence the exchange and transfer of ideas, expertise and knowledge.
- Government, through QRDI and/or QNRF, should establish a locally-relevant directorate to oversee, coordinate, facilitate, fund and continuously improve the IUGP activities—similar to IUCRC in the US or Fraunhofer Institutes in Germany. In addition, the directorate should host a database of local industrial problems and domestic universities' facilities and resources. It must consistently engage the university and industry actors in dialogue to form partnerships over the issues of common interest. Any partnership formed under the mandate of this directorate must receive government support and incentives. The directorate, which could be called as Qatar Industry-University Partnership (Q-IUP), should also have the mandate to integrate the local and traditional practices in Qatar's knowledge infrastructure.
- The suggested Q-IUP should have a priority mechanism for supporting innovative entrepreneurial activities by facilitating spin-offs out of the IUGPs.
- Qatar government needs to make major reforms to facilitate the development of private sector by extending the industry-government partnerships. Realizing the diversification of economy would not be possible without the development of private sector and non-governmental funding opportunities. In this regard, the government should further focus on: (1) encouraging independent financing schemes under legal framework to ensure secured private leasing; and (2) establishing effective credit registries and bureaus to facilitate the financing schemes. An increase in the number of deposit and savings banks, finance and leasing companies, money

lenders, and insurance corporations will ultimately increase the ease of doing business. This step will increase the FDI on one hand and the number of SMEs in the country on the other.

- Industries (including all business and economically relevant sectors, firms and organizations) should consider investing in local human capacity and R&D as part of their core business activities, and not a burden or corporate social responsibility. With the rapid technological development in the world and ever-changing knowledge landscape, sooner or later, investing in local human capacity and R&D will become a business necessity for organizations in Qatar; only those firms will be at an advantage which align their business model with local human capacity and needs.

- Industry should seek innovative and entrepreneurial outcomes out of its support with academia, such as to help establishing innovative small businesses as its suppliers, or to diverse the economy instead of acquiring solutions, products or services from outside the country.

- Universities, on other hand, need to reconsider and reposition their mission and vision to accommodate the changing needs of the modern society, economy, and industry; the universities' leadership need to be agile in reconfiguring their services. The academic leadership should responsively align their offerings with the local and future needs and should ensure a growing, well-equipped, knowledgeable, skillful and purposeful graduate workforce with capacities and capabilities to start contributing to the society while they are studying. Presently universities are slow, bureaucratic, and non-responsive to the needs of rapidly evolving business and industry outlook, which makes them less relevant to the economic diversification policies of the country.

- Universities should direct their human resources (faculty, scientists, and students) to closely work with industry on domestically relevant issues—not for the sole purpose of academic publications, but also for immediate, near-term, and tangible impacts. This should be aligned with changes in how universities currently evaluate the performance of faculty members, researchers, and students.

- The universities should focus on increasing the number and improving the quality and relevance of the graduate programs in Qatar. At the same time, the current programs and new offerings should be aligned with the industry needs in order to solve the local problems on one hand and to seek support from the industrial partners on the other. Also, the universities must have industry representation in their governing boards in order to steer the development of human resources in the domestically relevant directions.

- Since intermediaries are held accountable for supporting and funding early commercialization efforts of research outcomes, it is necessary to involve them at an early stage of research projects. In fact, the participation of intermediaries at the proposal stage should be encouraged by the national research funding agency. Intermediaries should get involved in the planning and execution of the research projects and should ensure a commercializable quality of research outcomes. Also, intermediaries must encourage and facilitate the faculty members and researchers in knowledge transfer and spinoffs.

- Intermediaries should also ensure the availability of adequate funding through various sources (seed, angel, venture capital, and crowd financed) for supporting entrepreneurial activities in the country. The funding should be sufficient to support the innovators from idea generation to scale up and commercialization. The locally developed products and services should be promoted in the international markets through government-supported procurement or risk financing schemes.

Some other general suggestions are as follows:

- The regulatory framework in the country needs a revamp to gain confidence of the global and local community. Regulations in areas, including but not limited to, contract enforcement, company registration, employment, free zones, and private investment should be improved to meet the international standards.
- There is a need to organically nurture a world-class university in Qatar as a driver for academic, business, and cultural transformation.
- The number of researchers and scientists through home-grown PhD programs, and the gross expenditure on R&D should be increased with suitably endorsed research facilities.
- Existing and future R&D infrastructure in universities should be made easily accessible to industries and universities in order to eliminate the cost of duplication and idleness.
- The recent proliferation of innovation and entrepreneurship plans in Qatar resulted in some overlapping of programs. Streamlining such programs and avoiding redundancies can make these more effective.

References

Ahmed, F. B. J. (2018). Challenges of the knowledge society: Exploring the case of qatar. *Global Economic Observer, 6*(1), 39–54.

Al-Mana, A. A. (2017). *Measuring efficiencies and value creation of national oil companies: A case study of qatar petroleum*. International School of Management.

Biernacki, P., & Waldorf, D. (1981). Snowball sampling: Problems and techniques of chain referral sampling. *Sociological Methods & Research, 10*(2), 141–163.

Board of Trade of Metropolitan Montreal. (2011). A look at Canadian University-industry collaboration. Retrieved from https://www.ccmm.ca/documents/activities_pdf/autres/2010_2011/ccmm_rdvs-savoir_2011_en.pdf.

Conventz, S., Thierstein, A., Wiedmann, F., & Salama, A. M. (2015). When the Oryx takes off: Doha a new rising knowledge hub in the Gulf region? *International Journal of Knowledge-Based Development, 6*(1), 65–82. https://doi.org/10.1504/IJKBD.2015.069443.

Edmondson, G., Valigra, L., Kenward, M., Hudson, R. L., Belfield, H., & Koekoek, P. (2012). *Making industry-University partnerships work: Lessons from successful collaborations*. Retrieved from https://www.sciencebusiness.net/sites/default/files/archive/Assets/94fe6d15-5432-4cf9-a656-633248e63541.pdf.

Fryer, R. G., Levitt, S. D., & List, J. A. (2008). Exploring the impact of financial incentives on stereotype threat: Evidence from a pilot study. *American Economic Review, 98*(2), 370–375. https://doi.org/10.1257/aer.98.2.370.

Gremm, J., Barth, J., Fietkiewicz, K. J., & Stock, W. G. (2018). *Transitioning towards a knowledge society: Qatar as a case study* (1st edn.). https://doi.org/10.1007/978-3-319-71195-9.

Gulf Research Center. (2017). *Demography, migration, and the labour market in Qatar*. Retrieved from http://gulfmigration.org/media/pubs/exno/GLMM_EN_2017_03.pdf.

Hall, B. H. (2004). University-Industry research partnerships in the United States. In *European University Institute working paper (No. 2004/14)*. Retrieved from http://cadmus.eui.eu/bitstream/handle/1814/1897/ECO2004-14.pdf.

Kaklauskas, A., Banaitis, A., Ferreira, F., Ferreira, J., Amaratunga, D., Lepkova, N., et al. (2018). An evaluation system for University-Industry partnership sustainability: Enhancing options for entrepreneurial universities. *Sustainability, 10*(2), 119. https://doi.org/10.3390/su10010119.

Kothandaraman, P., & Wilson, D. T. (2001). The future of competition: Value-creating networks. *Industrial Marketing Management, 30*(4), 379–389. https://doi.org/10.1016/S0019-8501(00)00152-8.

Lamine, C. (2005). Settling shared uncertainties: Local partnerships between producers and consumers. *Sociologia Ruralis, 45*(4), 324–345. https://doi.org/10.1111/j.1467-9523.2005.00308.x.

MacDonald, A., Clarke, A., Huang, L., & Seitanidi, M. (2019). Partner strategic capabilities for capturing value from sustainability-focused multi-stakeholder partnerships. *Sustainability, 11*(3), 557. https://doi.org/10.3390/su11030557.

McSparren, J., Besada, H., & Saravade, V. (2017). *Qatar's global investment strategy for diversification and security in the post-financial crisis era (No. 02/17/EN)*. Retrieved from https://socialsciences.uottawa.ca/governance/sites/socialsciences.uottawa.ca.governance/files/cog_research_paper_02_17_en.pdf.

Ministry of Economy and Commerce. (2016). *Investment in the State of Qatar* (p. 26). Retrieved from https://doha.msz.gov.pl/resource/66e6d9b5-af2d-42a8-b329-5cf4e5df4c5c:JCR.

Moeliodihardjo, B. Y., Soemardi, B. W., Brodjonegoro, S. S., & Hatakenaka, S. (2013). *Developing strategies for University, Industry, and Government partnership in Indonesia*. Retrieved from https://www.adb.org/sites/default/files/publication/176593/ino-strategies-uig-partnership.pdf.

Oxford Business Group. (2015). *Qatar eases restrictions on FDI to boost growth*. Retrieved from https://oxfordbusinessgroup.com/overview/qatar-eases-restrictions-fdi-boost-growth.

PwC. (2016). *Doing business in Qatar: A tax and legal guide*. Retrieved from https://www.pwc.com/m1/en/tax/documents/doing-business-guides/doing-business-guide-qatar.pdf.

Qatar News Agency. (2018, December 1). *Education city: A cornerstone of development, progress in Qatar. Gulf times*. Retrieved from https://www.gulf-times.com/story/614768/Education-City-A-cornerstone-of-development-progre.

Qatar University. (2018). *Qatar puts the pursuit of knowledge front and centre* (p. 16). Retrieved from https://www.nature.com/articles/d42473-018-00045-9.

Thier, M. (2017). Can Qatar buy sustainable educational excellence? In *Imagining the future of global education* (pp. 99–118). https://doi.org/10.4324/9781315108711-6.

Tuten, T. L., & Urban, D. J. (2001). An expanded model of business-to-business partnership formation and success. *Industrial Marketing Management, 30*(2), 149–164.

Vohra, A. (2019, January). *Qatar not only survived the GCC blockade, but also thrived. TRT world*. Retrieved from https://www.trtworld.com/magazine/qatar-not-only-survived-the-gcc-blockade-but-also-thrived-23082.

Weber, A. S. (2014). Education, development and sustainability in Qatar: A case study of economic and knowledge transformation in the Arabian Gulf. In A. W. Wiseman, N. H. Alromi, & S. Alshumrani (Eds.), *Education for a knowledge society in Arabian gulf countries* (pp. 59–82). Emerald Group Publishing Limited.

World Population Review. (2019). *Qatar population*. Retrieved April 25, 2019, from http://worldpopulationreview.com/countries/qatar-population/.

Subject Index

© Springer Nature Switzerland AG 2020
W. Nawaz and M. Koç, *Industry, University and Government Partnerships
for the Sustainable Development of Knowledge-Based Society*,
Management and Industrial Engineering, https://doi.org/10.1007/978-3-030-26799-5